Presented with Pleasure to
Cuddy Chicks
Maurice Smith
Nov. 14, 1995.

THE TRAIL BLAZERS OF CANADIAN AGRICULTURE

A History of Agricultural Pioneering
and Development in Canada

By Dr. Karl Rasmussen, FAIC

The Trail Blazers of Canadian Agriculture
by Dr. Karl Rasmussen, FAIC

Copyright 1995 by the Agricultural Institute of Canada Foundation.

All rights reserved. No part of this publication may be reproduced, stored in a retrieval system or transmitted in any form or by any means, electronic, mechanical, photocopying, recording or otherwise, without the prior written permission of the Agricultural Institute of Canada Foundation, Suite 907, 151 Slater Street, Ottawa, Ontario K1P 5H4 Canada.

Published by the Agricultural Institute of Canada Foundation, Ottawa, Ontario.

ISBN 0-9690453-6-0

Cover design: Love Printing.
Printed and bound in Canada.

Price in Canada: $25.00.
Price in other countries: $30.00.
Prices subject to change without notice.

Foreword

The earliest Europeans to arrive in what is now Canada, came for the purpose of harvesting the rich fish stocks and pursuing the fur trade. They were not interested in agriculture nor in establishing settlements. Nevertheless, there was an obvious need for local sources of food to bolster and supplement available food supplies.

Response to that need required courage, initiative and perseverance in the face of major adversities created by climate and primitive conditions.

The ancillary role that agriculture played in the early times may be the main reason why this phase of our agricultural history has never before been adequately chronicled. Farming or gardening projects were usually limited in scope and often transitory. They were frequently scattered over vast distances, between geographically isolated locations. Consequently, records of these endeavours, both failures and successes, were usually lodged in archives and frequently buried in reports of other activities and events.

The phenomenal growth of Canadian agriculture in modern times has occurred in little more than a century. It has been well documented. In contrast, the history of the pre-modern period of Canadian agricultural development covers three centuries. Because of the comparative obscurity of information about this period, and due to the slow growth in agricultural development, it received relatively little attention from writers and historians.

Yet the story which has been revealed as a result of Karl Rasmussen's perception and talent is a story rich in human interest and one of remarkable response to daunting challenges. It is the story of the men and women who struggled to convert forest, marshland, upland and prairie to productive farm land. It tells of remarkable accomplishments, reflecting the ingenuity, initiative and perseverance of the pioneers. It recognizes the contribution of our native peoples to the frequently sparse food supply of the settlers, through the availability of Indian corn.

The Agricultural Institute of Canada Foundation is indeed grateful to the family of the late Dr. Karl Rasmussen for the opportunity to publish his book entitled *The Trail Blazers of Canadian Agriculture*.

Dan Rose, P.Ag.
President, Agricultural Institute of Canada Foundation

Table of Contents

List of Tables iii
Preface v
Acknowledgements vii
Introduction 1
Chapter I: Newfoundland 7
   Settlement and Agriculture Discouraged 10
   Agricultural Society Formed 15
   Legislation Enacted 22
Chapter II: Nova Scotia 25
   Marshland Development 29
   Subsidies Instituted 34
   Practices Criticized 41
   Legislation Enacted 47
Chapter III: New Brunswick 53
   Timber was Dominant 61
   Floods, Fines and Pests 68
   Legislation Enacted 69
Chapter IV: Prince Edward Island 73
   Land Clearing an Arduous Task 76
   Absentee Owners 78
   Livestock Improvement 84
   Legislation Enacted 92
Chapter V: Quebec 95
   Policies Discouraged Farming 100
   Demonstration Farms 102
   Early Research Efforts 111
   Legislation Enacted 120

Chapter VI: Ontario 125
   Continuous Production in Southern Ontario 130
   Marketing a Challenge 140
   Government Support Programs 146
   Legislation Enacted 151
Chapter VII: Manitoba 155
   Early Livestock Production Efforts 158
   Government Regulations 169
   Settlement Encouraged 173
   Wheat to Europe 177
   Grasshopper Plagues 180
   Legislation Enacted 182
Chapter VIII: Saskatchewan 187
   Environmental Hazards 189
   Land Ownership a Problem 195
   Transportation Problems 199
   Grants to Societies 204
   Legislation Enacted 206
Chapter IX: Alberta 211
   Introduction of Livestock 217
   Native Hostility Delayed Southern Development 221
   Irrigation an Important Factor 227
Chapter X: British Columbia 233
   Expansion Throughout the Mainland 238
   Gold Rush Stimulated Agriculture 241
   Fraser Valley "Best in the World" 244
   Associations Help Farmers Learn 247
   Proclamations and Ordinances 251
REFERENCES 255
INDEX 293

List of Tables

Newfoundland

    Population and agricultural statistics for a number of years . . . 20

    Annual imports of live animals for selected years 21

Nova Scotia

    Census data for the Beaubassin area for three separate years . . 32

    Basic data from the 1707 census 32

    Population and agricultural statistics for certain areas
in Nova Scotia in 1782 39

    Periodic population and agricultural statistics for Acadia
and Nova Scotia during early development, with 1966 data
for comparison 52

New Brunswick

    Estimated yields of crops, in bushels per acre 71

    Periodic population and agricultural statistics up to$_*$
and including 1861 with 1966 data for comparison 72

Prince Edward Island

    Livestock census data for the years 1751, 1752, 1753 76

    Population and agricultural statistics for selected years 94

Quebec

    Population and agricultural statistics for the period
1667 to 1735 104

    Periodic population and agricultural statistics for
New France, Lower Canada and Quebec during early
development, with 1966 data for comparison 124

Ontario

    Population and agricultural statistics for Upper Canada
and Ontario 154

Manitoba

    Summary census statistics for the Red River Settlement 167

    Manitoba population and agricultural statistics 185

Saskatchewan
  Population and agricultural statistics for Saskatchewan . . . 209
Alberta
  Population and agricultural statistics for Alberta 232
British Columbia
  Population and agricultural statistics for British Columbia . . 254

Preface

The agricultural trail blazers in Canada, the east coast fishermen and the fur traders who penetrated the continent, have yet to find their rightful place in the history of this country. The purpose of this book is to attempt to correct this deficiency, to provide the story of pioneering agriculture, a phase of Canadian development that has received too little attention in the past.

The professional historian may well find much to criticize in the presentation of the material in this book. In the main it ignores wars and national and international politics, the backbone of formal histories. The format and the contents have been chosen to provide, as much as possible, an eyewitness account of the beginning and early development of agriculture, an industry that has become of major significance in the economy of Canada.

The book is a factual presentation with some limited analysis of factors affecting the early development. Some readers may feel that there is too much detailed description, but this seems to be necessary to provide a realistic picture of what actually occurred. Brief summaries would fail to convey fully the flavour of the times and the conditions under which the developments took place. The very earliest pioneers, the fishermen and the fur traders, were operating in unknown territory with no background information on soil conditions, climate and other factors on which successful agricultural production depends. They were the ones who provided the first information on what the agricultural potentials and the problems of production were for the various parts of Canada.

Acknowledgments

By the author:

The search for the information that constitutes the substance of this volume could not have been successful without the cooperation of numerous individuals and institutions, to all of whom I acknowledge my indebtedness. Special recognition is given to the privilege granted me of access to the Hudson's Bay Company archives, a very rich source of background information, especially for western Canada. The Public Archives of Canada, and of the provinces of Prince Edward Island, Nova Scotia, New Brunswick, Saskatchewan, and British Columbia, the National Library, the Ottawa Public Library, and the Library of Agriculture Canada were rich sources of material. At all times the staff of these institutions were most helpful and courteous in assisting me, an amateur in the field of historical research.

By the family:

The Rasmussen family thanks the Agricultural Institute of Canada Foundation for undertaking the publication and distribution of this book.

The family and the Foundation both wish to record their appreciation to a number of individuals who contributed significantly to this project, some of whom made monetary contributions. They wish to recognize Dr. August Johnson, P.Ag., and the late Dr. Wallace J. Pigden for initiating the project and supporting it throughout, and Alan Bentley, P.Ag., for steering the project to its successful conclusion.

Introduction

Agriculture is a major industry in Canada, an industry that had its beginning with the very earliest Europeans to touch its shores. The history of the development of this industry is still to be written. There have been some substantial contributions to the history in a couple of the provinces but for others this is lacking. On the whole, relatively little attention has been given to the real pioneers of agriculture in this country, the fishermen and fur traders who first tested the potential and experienced the hazards and limitations of this country for food production. It may be argued that much of this early activity had little or no major impact on later development and yet, it did provide a background of information about the agricultural potential or lack of potential in the different regions. In addition, in overcoming obstacles, it had its share of exploits that are worthy of recording and that add color to the early history of Canada.

Agriculture, as a significant industry in Canada, developed very slowly when compared to the situation in the United States. The main reason for this was the difference in the primary motivation for settlement in the two countries. In the colonies to the south the motivation was the desire to escape conditions in Europe and to make a permanent home and develop agriculture as a means of self-sufficiency. Fishing and fur trading became important industries and aided in the economic development of the colonies but were never the "raison d'être" that they were in Canada.

As far as Canada is concerned the first attraction to the east coast, other than the search for a route to the Orient, was primarily the rich fisheries that became known in the latter part of the 15th century and were exploited by the fishermen of several nations in the 16th century. This led to contact with the natives and soon trading with these people led to the development of the fur trade, which in turn became the primary interest in the interior of the country. Thus, both the French and English in Canada were more concerned with the development of these two industries than with the development of the country as a place for an increased population. Agricultural development was in direct conflict with the presence of fur-bearing animals. This was recognized by all concerned and the fur traders individually and the trading companies collectively were opposed to the development of an agricultural industry.

The fisheries off the east coast were seasonal, with the fishing fleets coming from Europe each spring and returning home in the fall. They carried provisions with them and were not interested in developing a Canadian-based food supply. It would seem that in a few cases an effort was made to produce a summer garden, for Hakluyt published a letter from an English fisherman which lists the ships fishing off Newfoundland in 1578 and in this he adds: "I have in sundry places sowen Wheate, Barlie, Rie, Oates, Beanes, Pease and seeds of herbs, kernels, Plumstones, nuts, all of which prospered as in England."[1]

As long as the fur trade was confined to the areas adjacent to the St. Lawrence River and Hudson Bay the problem of food supply was not insurmountable as supplies could be brought from Europe at reasonable cost and could be supplemented with the meat of wild animals and the fish from the rivers and lakes. But as the trading posts moved farther and farther inland the great distances and the primitive means of transportation increased the cost of imported food greatly and became a heavy drain on prospective profits. It then became an economic necessity to reduce these costs and the simplest way was to depend on locally-produced food. This continued to be fish and game until well into the late 18th and early 19th centuries but gradually locally-grown agricultural products became increasingly important.

Even before the trading posts moved inland, and while the Indians still were bringing their furs from distant points, they had problems in obtaining an adequate supply of food to sustain them during their travel. Innis outlines this as it developed after the Huron Indians had been driven from their homes in what is now western Ontario. "These Indians no longer had an adequate supply of corn with which they could support themselves in the prosecution of the trade. The fur trade devolved upon the French to an increasing extent and it became increasingly expensive. French traders, as in the case of Radisson and Groseilliers, found it necessary to penetrate to the more remote regions around Lake Michigan and Lake Superior to stimulate new interest in the fur trade and to encourage the growing of corn so that the longer journeys to the colony could be undertaken. According to Radisson, 'my brother stayed where he was welcome and putt up greate deale of Indian Corne that was given him. He intended to furnish the wildmen that weare to goe downe to the French if they had enough'."[2] The western and northern Indians trading at Hudson Bay had similar problems in food supply for their long journeys from inland points.

Cost of food was not the only factor involved. The Europeans were accustomed to a varied diet and were not satisfied to subsist on fish and meat to the exclusion of vegetable and cereal products. To overcome the problem, at least in part, production of food at the trading post was the alternative, and eventually this became an integral part of the trading program. The need was accentuated by the fact that game and fish often were an inadequate source of food. In many cases the scarcity of these products left both the Indians and the traders face to face with starva-

Introduction

tion. One finds in the Hudson's Bay Company post journals repeated references to such situations.

The story of the transport system of the fur trade has been well told and will not be repeated here. However, certain aspects of direct interest to this story have not been told. Transportation of agricultural items, particularly domestic animals, presented a real challenge in view of the primitive means of transport and the distances to be covered. Even the Atlantic crossing, which sometimes dragged out two or three months, presented major difficulties, not the least of which was the lack of an adequate water supply. While rain could be expected to provide some replenishment *en route* this did not always suffice. One report stated that water ran short and it was necessary to give the goats and swine cider to keep them alive.

During the first century or more of the fur trade, the transport of material and personnel to the interior and the transport of furs out was by canoe in both east and west. Later, York boats became important in the west and scows and ships were used in the east. The canoe presented no major impediment to the transport of seeds or potatoes but it was no simple matter to transport poultry and domestic animals in canoes and yet these were transported hundreds of miles through the wilderness. This will be dealt with in more detail at the appropriate time.

In determining the earliest agricultural activities it is necessary to define the term agriculture. In its simplest terms, and the term appropriate to the time under consideration, agriculture is: "The production of foodstuffs necessary for the sustenance of people" or, more specifically, "The cultivation of plants and domestic animals to provide the major source of food for a non-nomadic population." In this there is no limitation imposed on the size of the operation. In North America we are accustomed to think of agriculture as commercial agriculture, but there still remain large parts of the world where subsistence agriculture prevails in the same way that it did in the early part of the period that we are considering.

The other factor that must be defined is the time frame in which we are interested. While there was agricultural development in various parts of what is now Canada before the advent of the Europeans, our concern is with the introduction of agriculture by them. The purpose is to identify its origin, in what are now the various provinces, and to trace the development to the point where agriculture was established as an important industry. Some effort will be made to indicate the differences in the rate of development in the various areas but the changes are too great to be discussed in detail in this work. Furthermore, the primary interest is in the establishment not the ultimate development.

Though the agriculture practised by the Native Peoples will not be of major concern it seems pertinent to point out that there were important crops grown by the Natives of eastern Canada and these became of

The Trail Blazers of Canadian Agriculture

significance to the Europeans. The main crops were maize, or Indian corn, pumpkins, squash, beans, and tobacco. In the Canadian prairie regions there was essentially no agriculture, though there is a suggestion of some activity in southern Manitoba. Though the Indians grew tobacco it is interesting that tobacco became one of the prime articles of trade; but it was tobacco raised in the south and processed in Europe that entered into the trade. Indian corn became a crop of great importance to the early settlers, particularly in eastern Canada, but most of the early agriculture of Europeans depended on the crops that were introduced from Europe and with which they were familiar. Another factor was that much of the early settlement was in areas where the climate was not favourable for corn production.

In determining the criteria that would be used in selecting items for inclusion in this review it was decided that continuation of agricultural activity at a particular site should not be a limiting factor. The only factor used was a definite, acceptable record of actual crop production. This means that in some cases there are records of production in areas where there is no real agricultural production today. But it was thought to be desirable to use this base in order to place on record the actual first beginnings of planned food production in various parts of Canada, now delineated as provinces.

Some of the earliest recorded cropping activities were haphazard and relatively unplanned and are of only peripheral interest. It seems that it was not uncommon practice for the early explorers and the fishermen frequenting the coast of Canada to carry with them seeds of various crops and to plant them from time to time. For example, we read that Frobisher, on his third voyage of exploration in 1578, had planned to leave a wintering party on one of the arctic islands. While this did not materialize a stone house was built "upon the Countesse of Warwickes Island, to the ende we might prove against nexte yeere, whither the snowe could overwhelme it, the frosts breake it up, or the people dismember the same. Also we sowed pease, corne, and other graine, to prove the fruitfulnesse of the soyle against the next yeere."[3]

Many of the earliest efforts at agriculture occurred in those parts of Canada that are outside of the areas of commercial agriculture today. They were undertaken where the soil and climate were not conducive to success in the growing of most crops. The early efforts were made by fur traders whose trading success depended on the presence of beaver and other fur-bearing animals in great numbers and, generally, these animals were found most abundantly in the areas least suitable for agriculture.

The national politics, and even the intrigues, that affected the early trading companies and the international rivalries and wars that influenced development will not be given direct consideration or discussion. Suffice it to say that it is recognized that these activities had a direct

Introduction

impact on early development. Later on, international politics became important in relation to markets for agricultural products.

Having now set the background and outlined the criteria that will be used, it is proposed to outline the earliest developments in each province. These areas will be dealt with in geographical sequence rather than in a strictly chronological order.

The Trail Blazers of Canadian Agriculture

Chapter I
Newfoundland

Newfoundland, the latest province to become a part of Canada, is the province with the distinction of first having agricultural activity practised by white settlers. However, it lagged behind the other eastern provinces in reaching the stage where agriculture was recognized as an important industry that warranted attention and support from the government. Agricultural development grew out of the fur trade in most parts of Canada, but the fisheries were the forerunners in Newfoundland.

Despite continuing argument about the location of Vinland and the landfall of the early Norse explorers, there is strong evidence that Norsemen, led by Thorfinn Karlsefni, established a settlement at L'Anse aux Meadows on the northern tip of Newfoundland about 1008. According to Morison this colony of 250 men and women brought with them all kinds of live stock and cleared land for cultivation.[1] For various reasons the venture was not successful and did not survive more than a few years.

After the departure of the Norsemen, active European contact with Newfoundland lapsed until late in the 15th and early in the 16th centuries when fishermen, coming to the Grand Banks for fish, began to use Newfoundland as a summer base for their activities. It is difficult to determine with certainty when actual settlement first occurred. Prowse states: "We cannot fix the precise year in which settlement commenced in Newfoundland; Sabine states that in 1522 there were 40 or 50 houses in Newfoundland; but from scattered information contained in the records, and narrative of voyages, it is clear that, from a very early period, a few crews were left behind every winter to cut timber for building cook-rooms, stages, train vats, wharves and for the construction of boats."[2]

Information cited by Biggar[3] indicates that there was a Portuguese settlement in Newfoundland in the period 1520-25 but there is no specific information to indicate that any agriculture was practised.

The first record of crop production in the province comes from Edward Haie (or Hays): "The soyle along the coast is not deepe of earth, bringing forth abundantly peason, small yet good food for cattel. Roses, passing sweet, like unto our muske roses in forme, raspases, a berry called harts, good and holesome to eat. The grasse and herbe doth fat sheepe in very short space, proved by English merchants which have carried sheepe

The Trail Blazers of Canadian Agriculture

thither for fresh victuall, and had them raised exceeding fat in lesse than three weeks. Peason which our contrey-men have sowen in the time of May, have come up faire, and bene gathered in the beginning of August, of which our Generall had a present acceptable for rareness, being the first fruits coming up by art and industrie, in the desolate and dishabited land."[4] This was written in reference to Gilbert's expedition in 1583, which was a complete failure.

The first real effort at establishing a permanent settlement with an agricultural base came in 1610 when a charter was granted to "The treasurer and the Company of Adventurers and Planters of the City of London and Bristol for the Colony of Plantations in Newfoundland." In this same year John Guy, the first governor, led out the first group of settlers, to the number of 40 or so individuals. He established his colony at Cuper's Cove (Port de Grave) on Conception Bay and that year built a fort, a dwelling house, a workshop, and a boat, sowed grain, and in general made preparations for the winter.[5]

Among the instructions given to Guy, the following are of particular interest: "As the benefit and use of domestic creatures without whom not onlie any desolate countries but also a (civilized countrie) could not well be inhabited we would have you take with you a small number of evrie kind of them male and female water and other things needful for their transportation not (placing them on the sea shore) but either by islands or necks of lands between Bayes where together with the care of herdsman they shall escape from wild beasts and so increase and multiplie of which number (we would) not have anyone killed without great and urgent cause."[6]

Apparently Guy did not follow instructions though he did take some live stock with him. This is evident from a letter that he wrote on May 16, 1611: "We have sowed all sorts of graine this spring, which prosper well hitherto. Our goats have lived here all this winter; and there is one lustie kiddie, which was yeaned in the dead of winter. Our swine prosper. Pidgens and cowies will endure exceedingly well."

"Our poultry have not onely laid eggs plentifully, but there are eighteen young chickens, that are a week old, besides others that are hatching....For as concerning sending of cattle, it will be best that it be deferred untill next spring."[7]

In that year additional settlers arrived and it is reported that: "on Southern River they erected mills, houses, and farm buildings; a considerable quantity of land was cleared and surrounded by stone walls."[8] Guy stayed through 1610 and 1611 but late in that year he returned to England. However, the following year he returned with new settlers and horses, cattle, pigs, poultry, and farming implements.[9]

Additional attempts at colonization were undertaken by several prominent men in England but they did not seem to have had any real success. The first colony that did survive for a while was that of Lord Baltimore in 1621. In that year he opened a settlement at Ferryland under the leadership of Captain Wynne. Some indication of the early progress of this settlement is given by Wynne in letters to Lord Baltimore, who at this time was still in England. On July 28, 1622, Wynne wrote, outlining what had been accomplished. After detailing the construction that had taken place he continued: "We got also together as much firewood as will serve us yet these months. Wee also fitted much garden ground for seed, I meane Barley, Oates, Pease, and Beanes."

"Notwithstanding this great task for so few hands, we have both Wheat, Barley, Oates, Pease and Beanes about the quantity of two Acres. Of Garden roome about halfe an Acre: the corne though late sowne is now earing, the Beanes and the goodliest Pease I ever saw, have flourished in their bloomes this twenty dayes. Wee have a plentifull kitchin garden of Lettice, Radish, Carrets, Coleworts, Turneps and many other things. We have also at this present, a flourishing medow of at least three Acres, with many hay-cocks of exceeding good hay, and hope to fit a great deale more against another yeere."[10]

A later letter expanded on this report with considerable apparent enthusiasm. "We have also a plentifull Kitchin-Garden of many things, and so ranke, that I have not seen the like in England. Our Beanes are exceeding good: our Pease shall go without compare; For they are in some places as high as a man of extraordinary stature; Raddish as bigge as mine arms: Lettice, Cale or Cabbedge, Turneps, Carrets and all the rest is of like goodnesse. We have a medow of about three Acres: it flourished lately with many cockes of good hay, and now is made up for Winter feeding. We hope to be well fitted with many Acres of medow against another yeare: of pasture land, we have already to serve at least three hundred head of Cattell; And to all of this, if it please God, a good quantity of feeding ground shall be fitted, and such buildings as we shall be able to accomplish."[11]

Wynne continued: "Touching the soile, I find it in many places, of goodnesse far beyond my expectation: the earth as good as can be: the Grasse both fat and Unctious, and if there were store of Cattel to feed it up, and with good ordering, it would become a most stedfast nourishment: whereof the large breed of Cattel to our Northerne Plantation, have lately given proofes sufficient, though since, they have become shamefully destroyed."

"At the Bristow Plantation, there is as goodly Rye now growing as can be in any part of England; they are also furnished with Swine, and a large breed of Goates, fairer by far than those that were sent over at first. The Vines that came from Plimouth, doe prosper well: nay, it is to be assured, that any thing that growes in England, will grow and prosper verie well here."[12]

Captain David Powell, who brought supplies in 1622, mentioned an area of a thousand acres of good ground on a peninsula, with a narrow neck, where cattle could be kept secure from predators.[13]

Settlement and Agriculture Discouraged

Despite the apparent success of some of these early settlements, hard times were ahead as the English merchants and fishermen did everything in their power to hinder the settlers and prevent any additional development. As stated by Birkenhead, of the situation that had developed by 1633: "At this time the resident population of the island cannot have exceeded a few hundreds, and every step was adopted which a vicious political economy could suggest to keep the numbers down. It was penal for a settler to dwell within six miles from shore, for any planter or inhabitant to take up the best positions in the harbours before the arrival of the fishing fleet in the spring; and every master who sailed with a crew to Newfoundland was under bond — lest here and there a permanent settler should filter through — to return with his exact complement of Hands."[14]

Despite the restrictions, some development took place. In 1676 an attempt was made to put a complete stop to this, for by an "order of January 27, 1676, all the Newfoundland settlers were called upon to relinquish their property, to surrender at a moment's notice the patient labour of a lifetime."[15] Through intervention by leading men the order was rescinded but the lot of the settlers did not ease greatly.

An indication of the limited extent of development is provided, in 1677, by the Commodore of the convoy, who served as governor during the summer and made the following report:

"A list of Inhabitants and their concernes from Trepassey to Cape Bonavista

House keepers	162
Wifes	64
Sons	137
Daughters	130
Men servants	1327
Women "	15
Horses	12
Cattle	480
Sheep	131
Hoggs	845
Gardens	114."[16]

From this list it is evident that agricultural activity was quite limited, although there is no specific information on the extent of the 114 gardens. However, there is corroboration on the limited nature of production in the periodic reports of the Commodores or Governors to the Board of Trade and Plantations. One of the questions regularly requiring to be answered was directed at the activities of the resident population and their source of sustenance. The answers differed from time to time, possibly depending on the emphasis the Commodore wished to make. Thus, in 1706, the following: "Their trade and manner of living is wholly by fishing, having no husbandry, nor Tillage, nor any Cattle, but what are brought thither from New England, Every Year."[17]

In 1708 the answer was slightly different, indicating considerable dependence on local production: "The Planters and Inhabitants have no other sustenance than what they get out of their own gardens."[18]

In 1711, in making reference to the threat of the French, the Commodore wrote: "makes that (i.e. trapping) much neglected, as it does many other Improvements the Country is capable of, as Building, Breeding of Cattle, planting many European Grains, Fruit Plants etc., necessary for their more comfortable Subsistence."[19] He apparently overlooked the fact that the greatest source of discouragement was not from the French but from their own countrymen.

Much later, in Governor Shuldhams's account of the fisheries in 1772-73, we find a fuller statement, including his evaluation of the agricultural potential. "The inhabitants are subsisted by Provisions brought from Great Britain, Ireland and the American Colonies, as there is little Land improved, or indeed capable of Culture, and what is, must be improved at very great Expense, perhaps at more than the Value of it, where the Soil is so ungrateful, and the Price of Labour so high. They breed some Swine, but for their Sheep and Cows, Bread, Flour, Pease, Rum, Sugar and Molasses they are obliged to America, as they are to Ireland for their Beef, Pork and Butter, and to Great Britain for a small mixture of these and all other necessaries of Life. Some of the inhabitants raise a few Poultry and provide Forage for fattening their Sheep, Oxen & Cows for their own private Consumption in the Winter, but all this with a great deal of difficulty & Expense."[20]

"The only objection that can be offered against the Boat Fishery is the Settlements that must be made in this Country to carry it on, and that instead of a Fishery it becomes a Colony, and that Great Britain loses the Annual returns of her Subjects, but this Objection can have no weight if we consider a certain fact, that no one can or has settled in Newfoundland but on account of the Fishery, for the soil is poor and Barren, nor can it be cultivated but at an immense Expense or granted but for the use of curing Fish, which is a Bar sufficient to exclude any schemer or a Planter."[21] It is clear that the Governor had a low regard for the agricultural prospects of Newfoundland. It is equally clear that he was voicing the conventional bias against the planters.

Very little detail is available about the cultural practices used during this whole early period. One may deduce, from oblique references and descriptions for later periods, that much of the cultivated land was tilled by hand and, in general, cultural practices were primitive, though ploughs and harrows were available. Some insight into conditions is given by Aaron Thomas, a seaman who visited there in 1794-95. In one case he describes an incident where a large number of lobsters were being boiled and he and others were cracking and picking them outside the door of the house. "No sooner had we begun picking the Fish than we were surrounded by Fowl, Cows, Ducks, Goats, Geese, Cats, Horses, Calves, Pigs, tame Sea Gulls, Sheep and Dogs, in short all the Specie of every kind which is domestic in Newfoundland. When I threw the Claw of a Lobster a Cat, a Goat, and a Hog would start after it, generally the Cat was Successful. When the Body of a Fish was checked off the Fowls, the Cows, and the Gulls and the Sheep would join in pursuit. I saw a Horse, a Calf, and a Dog smelling the same flesh, each in perfect good humour."

"The Shores of Newfoundland being the grandest mart in the known world for Fish it is the cheapest food in the Country. The Inhabitants give it to their domestic animals and they all eat it."[22]

On the 10th of June 1794 he noted: "Altho this is the middle of Summer in England, yet we had here last night a severe Frost, so intense that had it been with you — farewell to all your Fruits and Vegetables. The weather this morning as cold as ever I felt it in January in Britain. As soon as the Sun makes its appearance the cold is dissipated and its luminous rays are grateful and pleasing. This time of year is called Spring in Newfoundland. The Vegetable Kingdom is just peeping from its winter habitation and the cultivated spots are assuming their livery of green; the sun instills his genial qualitys into plants and herbs in a very bountiful manner here, they grow more in one week then they do in England in three. The summer being short, the Sun is very powerfull and everything is brought to perfection in a short space of time."[23]

Later he wrote: "The very high price of the Flesh and Fowl in Newfoundland is owing to neither Wheat or Barley growing here. I saw Four acres of Oats in good condition, near Titti Whitti Lake, which I believe was raised in the District of St. John's. An acre of such Meadow Land as environs the Castle of Kenilworth in Warwickshire would be a pretty little estate here. Very little Hay is grown here. I have paid Nine Pounds a Ton for Hay for some Stock we had on board the Boston. For the Summer months Cattle, Horses and Sheep fare well, but from the small quantity of Hay which can be housed little stock of Cattle etc. can be supported during the long winters in this Island. From this circumstance ariseth the high price of Fresh Provisions. Live Stock are brought to this place from Nova Scotia and the States of America and sold at reasonable rates, but they fall into the hands of the Merchants and the Jobbers who retail them out at exorbitant prices to the Fishermen and inferior class of Inhabitants."[24] This last complaint has a very familiar ring.

Early in the 19th century a shift in official policy began to be evident and in 1813 the first grants of land for agricultural purposes were made.[25] But a land tax or rental was imposed which in no way encouraged settlers to clear and cultivate farms. The tax was abolished in 1822 and it was at this time that agriculture began to get its first bit of real encouragement from government. Two hundred years had elapsed between the time of the first English agricultural settlement and the time when official encouragement of agriculture occurred.

The methods and conditions of disposal of crown land varied from time to time. For example, in 1830, if the grant was less than 20 acres, one third was to be cleared and cultivated within three years. If the grant exceeded 20 acres, one fifth was to be cleared and cultivated within the same time period. After that the grantee was free to use his own discretion as to the time and extent of its cultivation.[26]

Listings of persons petitioning for land under this policy show that a substantial number of grants were made. They varied in size from one to 150 acres with most of them in the range of 15 to 25 acres.

Three years later a change was made from a policy of grants to a policy of sales of lands beyond four miles of St. John's. The following were the terms under which purchases could be made:

4 to 6 miles from St. John's	7 shillings	6 pence per acre
6 to 8 " " "	6	0
Within 2 miles of the Towns of Harbor-Grace and Carbonear	5	0
All the rest of the Island	2	0

Not more than 20 acres per individual could be sold within two miles of Harbor-Grace and Carbonear and one mile of other principal settlements.[27]

In 1840 a further change was made and all petitions for land were to be approved by the governor. If approved, the land was to be advertised for sale and then put up for public auction at an upset price of 2 shillings per acre and sold to the highest bidder.[28]

While most of the land holdings were relatively small there were some fairly substantial holdings. Even before land could be held legally some holdings were developed. For example, on May 24, 1810, the following advertisement appeared in the *Royal Gazette*:

"For Sale: That fine situation at the North side of Placentia harbour, formerly occupied by Mr. John Power, containing a Dwelling house in the centre of three Gardens, well stocked with a variety of fruit trees; with a good Barn and Cooperage, as well as two stores, one two hundred feet long, and about twenty feet wide, the other between forty five feet long and eighteen feet wide, both two stories high: is about fifty or sixty acres of Land in a state of cultivation."[29]

Another advertisement, on October 15, 1833, of property to be sold to close a bankruptcy at Brigus, listed several farms: "Mount Dorset Farm, consisting of 80 acres of rich and highly cultivated Land, with extensive Erections thereon. The south Mount Pleasant Farm consisting of 60 Acres of excellent Land, chiefly under cultivation, with several valuable erections. The Celebrated Farm, Cochrane Dale, comprising Goold's Farm of about 250 Acres of Tillage, Meadow, and Woodland, with many valuable buildings thereon. Juniper Stump Farm, Consisting of about 150 Acres excellent Land, about 20 of which are under cultivation."[30]

The equipment available and in use may be inferred from advertisements of auction sales. Thus, July 19, 1810, a notice of auction included: "Stock, Farming Utensils, etc.4 Milch Cows, 2 Calves, 2 Pigs, 2 Horses, 6 Carts, 3 Cattamarans, 2 Wheel Barrows, 2 Ploughs, 1 Harrow, 2 Pitchforks, 3 Mattocks, 2 Pick-axes, 1 Crow-bar, 3 Shovels, 3 Garden Spades. About 1000 Firewood, 2 Sleighs, 2 Gigs. A quantity of Old Iron, About 1 Ton Hay, 1 Crosscut saw, 1 Hand ditto, 1 Jack, 1 Table, 1 Corn Bin, 3 Sets Cart Tackling, 2 Saddles, 2 Bridles, 1 large Boiler."[31] In the advertisement reference also was made to an orchard as part of the property.

On November 14, 1811, another auction was advertised with the following items: "1 Horse, 2 Cows, 1 Bull, 2 Heifers in Calf, 2 Ploughs, 3 Catamarans, 1 Harrow, 2 Carts, 2 Wheelbarrows, Some Tools, 5 or 6 Tons Hay, 2 Saddles and several Bridles, Husbandry Harness, a few m. Fish Cask Staves, Some Casks; a quantity of Potatoes, Turkeys, Ducks, Fowls, a large quantity."[32]

That there was no rapid change in equipment is indicated by the equipment listed for sale at auction in 1828: "Scotch ploughs, Harrows, a large Roller and Box Cart (complete), a Box Cart and Wheels, A Wood ditto ditto, Wheel Barrows, Hay and Garden Rakes, Spades, Shovels, Dung Forks, Pitch Forks, Hoes, Mattocks, Scythes, Crowbars, Mauls, Hatchets, Pick Axes, etc., etc.."[33]

One of the early indications of a change in official policy with regard to agriculture is found in a statement addressed to the Colonial Department by Chief Justice Forbes under date of August 14, 1822: "It were desirable that, with the view of opening some auxilliary employment to the inhabitants of Newfoundland, every restraint upon the cultivation of the soil should be removed, and every encouragement given to the breeding of sheep, cattle, and other live stock."

"The necessity of cultivating the soil, as an auxilliary to the fishery, is not disputed, nor is there any existing law which prohibits it, but there is none to encourage it, and there is still maintained in the island an ancient opinion, that it is against the policy of Government — as if that could be called policy, which, in a country overstocked with people and distressed for food, would prohibit so plain a dictate of natural law, as that of raising subsistence from the earth."

"This cannot be, is not the policy of the British Government, and nothing is wanted but a fair apprehension of the case to induce its enlightened rulers, not only to remove every shadow of obstruction from the cultivation of the soil, but to encourage and protect it by every means in their power."[34]

In 1825 Governor Cochrane inaugurated an active policy for the encouragement of agriculture as he believed that agriculture should be an important factor in the development of the colony. In furtherance of this, grants were made for the construction of roads to areas where there was agricultural potential. The effect was recorded by Bonnycastle in 1842. "The great drawback to agricultural pursuits is, however, the want of adequate manure and of roads. If there were roads, of course the miserable, half starved dogs which now draw the small farmer's supplies of wood would give way to horses, and horse-manure would be attainable."

"As soon as Sir Thomas Cochrane had opened communications by land with Conception Bay, by Portugal Cove and Topsail, before which time a horse had been an object of great novelty, even at the capital, and could only be supported at great expense by the imported hay and oats, fields soon were cleared and sown, and every merchant, and many fishermen, supplied themselves with horses for pleasure, for their agricultural pursuits, or for more easy communication with the capital of the colony."[35]

Despite the fine words of Justice Forbes and the encouragement by Governor Cochrane there continued to be a general reluctance to accept agriculture as an industry that could be of significance to Newfoundland. Thus, as late as April 8, 1834, in an Address to the King, the Legislative Assembly wrote: "Newfoundland is, and must ever remain, a mere fishing station attached to whatever country may be enabled, — by its own power, to keep it from falling into the hands of any other; — that it may always continue to be one of the numerous sources of wealth as well as maritime greatness of our parent state, is the earnest desire of every inhabitant of the Island."

"Agriculture can under no possible circumstances ever become more than an auxilliary of the fishery. It can never subsist as an independent branch of national industry, and the utmost limit to which it can be carried, will never exceed a precarious growth of vegetables and grass, with occasionally some coarse grains to eke out a scanty supply of provender for the small stock of cattle which it may be necessary to keep during the winter. For bread, and all other necessaries of life, the inhabitants of Newfoundland will ever be dependent on a foreign supply. It is therefore obvious that her population can never be reasonably expected to increase far beyond what the fisheries will afford employment for."[36]

Agricultural Society Formed

But change was taking place, albeit slowly, and on January 14, 1842, an Agricultural Society was formed at St. John's for the purpose of furthering the improvement of agriculture. It is interesting to note some of the comments of the new governor, Sir John Harvey, who spoke on this occasion: "It was, therefore, with equal surprise and plesure, that I discovered, as well from my own observation as from information of others, that Newfoundland is, in reality something more than a mere 'fishing station', and possesses resources beyond the mere 'rocks on which to dry the nets of the fishermen'. In a word, I saw in it the undoubted evidence of a capability for agricultural pursuits, far beyond what I had imagined to exist."[37]

"That this island, throughout almost the whole extent of its bays, harbours, and inlets, is skirted by a belt of cultivated land, varying in depth, from one to several leagues, well calculated to reward the labours of the agriculturists, of which no more convincing proof can be required than the specimens of produce now before you, consisting of wheat, barley, oats, turnips, potatoes, etc. equal in size, in weight, and in quality, to the productions of any country, England not excepted."[38]

"Gentlemen — My object in presenting myself here this day, is to make this declaration of my sentiment, in regard to the objects of the association now about to be formed, for the promotion of agriculture in this island, and thus publicly, to pledge to it every degree of encouragement and support which my position, as the representative of the sovereign, may enable to extend to, or procure for it."[39]

Further evidence of the change of attitude is noted in the Governor's speech to the Assembly, December 14, 1848, in which he said: "The subject that has mainly occupied my attention has been the extension of the field of labour for our increasing population, the fisheries alone being now inadequate for their support; and the promotion of Agriculture appears to me to be not only the surest but the sole course now remaining for us in the attainment of this end; for the increased cultivation of the soil will not only carry with it the germs of future comfort and happiness to our population, but will directly contribute, as a most powerful auxiliary, to the great staple Trade of the Colony, the Fisheries, which first laid the foundation of its commercial greatness, and must ever be the source of its prosperity."[40]

The Assembly, in its reply to this speech, echoed the sentiments of the governor, indicating agreement with this new attitude.[41] This was in sharp contrast to an assembly address to the King 14 years earlier.

At this point it may be desirable to look briefly at the general state of agriculture in Newfoundland at this time. Bonnycastle provides a general picture of some of the production practices that were being used. "The principal objects of agricultural industry are potatoes, oats, barley, hay, straw, turnips, and cabbages with the common garden esculent vegetables, which all thrive well, excepting onions, and they are imported cheaper than they could be reared at present, for want of proper manure."

"Wheat is growing now within a mile of the house I am writing in; it was sown in the fall of the year, and in this month, April, has survived all the severe alternations of the winter. Winter wheat, in fact, is better adapted to the climate than any other; as this grain, if sown in the spring, is apt to rot before it shoots, and the short summer will not allow a sufficient time for its growth."

"I do not believe that the sea-coast, covered as it is occasionally with saline vapours, will ever permit the vicinity of St. John's, or any of the harbours on the eastern or southern coasts, to become a wheat-growing region."[42]

"For the district of St. John's especially is well adapted as a grazing country, and the imported and home-raised cattle look as plump and as sleek as those of any other part of the world where they are carefully attended to, and I have seen cows at some of the farms which would not discredit the dairies of Devon."

"In my country walks I have also met, more than once a hundred head of stout beeves going a-field; and looking, from the eminences surrounding the city, over a country which lies spread out below you like a map, the numerous fields appear dotted with oxen and cows, as well as sheep; and, as before observed, it now only requires roads, to open more grass ground, and to supply sufficient fodder for the long and severe winter, to raise a stock adequate to the increasing demand, particularly as turnips, beets, and oats may be obtained in any quantity; and it is well known that on the south shore, numerous herds of cattle are reared, for even in the dreadful Bay of St. Shott's, the grave of the mariner, a solitary settler there has sixty head of neat cattle, for which, through want of roads, he has no market."

"The manure used in this country is the refuse of the seal and cod fishery, the garbage of the seal, and the heads of the cod, mixed with the absorbent, black, decomposing, exposed particles of the innumerable shallow peat-bogs, of the island, and with a reddish loam-looking earth, in which there is not a particle of lime, but much iron."[43]

"Some of the richer class here are using lime, both in their compost heaps, and on their fields, but as it is all imported, either from Spain or Great Britain, it is too serious an article of expenditure at present to do more than to enable them to test its value."

"Kelp is also used, and the sea-mud; but neither is in sufficient quantity near the capital, as the shores are high, bold and rocky, and afford no beaches of deposit."[44]

"The season for working on the east and south shores commences in May, and farmers house their last harvest late in October, or early November; for after the first week in the latter month strong frosts occasionally occur, and if the potato crop is neglected it is subject to be nipped in the ground, which renders the root bitter when placed in a warm cellar."

"They usually dig the potatoes all at once, as fast as possible, and make long narrow and shallow holes in the ground, upon which they heap the roots to a height of two or three feet above the level of the natural soil, forming them like a long pile of shot, with a ridge; which is then covered with the haulm or stalk of the plant, and that again with earth to some height. This is done to cause the potato to sweat, or dry, before it is again removed into the warm cellar or root-house. Sometimes these heaps, a little more covered and more carefully secured, are left all winter abroad, to be used as seed or food in the spring; for the frost does not enter the ground deeply in this country, and the severe winters are very rare."[45]

The 1848 annual report of the Agricultural Society adds some information regarding the livestock situation: "The importation which his Excellency has made of a bull and cow of the Ayrshire breed, will, it is hoped, be of ultimate utility, although it has not hitherto been the practice to breed and rear cattle in this district, principally owing to the large quantity of lean cattle imported from Prince Edward Island and Nova Scotia. Those persons, however, who had bred and reared cattle in the neighbourhood of St. John's as well as in other parts of the Colony, agree in the opinion that cattle bred in this Island are much hardier and better suited to the country than those imported. In the south and west parts of this colony, large numbers of cattle are bred and reared, and in the district of Placentia and St. Mary's alone, there are nearly 2,000 head of horned cattle kept, and about 800 in the district of Burin, and these numbers may not be considered the average stock constantly on hand. Hence it is obvious that if due attention were paid to the improvement of the breed, and if a system of agriculture were introduced and applied in those places where cattle are kept extensively, great advantage would arise, as well to the individuals as to the Colony, in a much larger amount of produce. The completion of one of the roads to Placentia would also be a great and most desirable aid in furtherance of these objects, by enabling cattle to be driven to St. John's at all seasons, especially in the spring when meat is scarce and dear, and before importations commence from neighbouring colonies. The premium which His Excellency the Governor has recently offered to encourage a better system of stall-feeding cattle will, it may be confidently expected, awaken attention to the object, and, tend to an improved practice in that department of agriculture."[46]

The 1849 report added a little additional information: "Our wheat is found to weigh, with few exceptions, not less than sixty pounds to the bushel, and our oats and barley maintain a proportionate character. With a view to encourage this important branch of our agriculture, His Excellency has caused several thousand bushels of seed corn, of the best and most suitable description, to be imported from various places, and distributed among the farmers throughout the colony. The result has been highly advantageous; the harvest has yielded fair return, and due care has been taken to mark and ascertain the varieties of seed which appeared to be best suited to our soil and climate."[47]

Part of the increased emphasis on grain was brought about by the regular occurrence of partial or complete failure of the potato crop, which had been the mainstay of the local food supply. The cause of these failures is not specified but reference is made, from to time, to disease and it may be assumed that it was blight that was the cause, the same disease that was creating havoc with the potato crop in other parts of the world at that time. The first specific reference to crop failure was in the minutes of the Council, January 4, 1833, when the need to provide seed potatoes also was discussed.[48] Additional references to crop failures were noted for the years 1838, 1842, 1845, 1847, and 1853.

New interest in agriculture was also indicated by the fact that prizes were given by the governor to stimulate production. The list of prize winners reported by the Agricultural Society in 1848 gives a fair picture of the emphasis and also of the small size of the acreages involved. A partial list is as follows:

- The Le Marchand Cup — for the greatest breadth of wheat crops, fair marketable quality for two consecutive years - to Hon. Patrick Morris.
- For the best crop of wheat, on any land of not less extent than three acres, £6 — to Mr. Michael Allen.
- For the best crop of wheat, on any land of not less extent than two acres, £4 — to Mr. William Hect, Grove Farm.
- For the best cultivated crop of wheat, on any land of not less extent than one acre, £3 — to Mr. John Harding, White Hills.
- For the best cultivated crop of wheat, on any land of not less extent than half an acre, £2 — Mr. James Shea, near Bally Holly.
- For the best cultivated crop of oats, on any land of not less extent than two acres, £3 — to Mr. John Droyer, Oaks Farm.[49]

The governor also encouraged livestock production with prizes for the first livestock show in Newfoundland, as noted in the 1849 report of the Agricultural Society. But this was not the first exhibition of produce of the land. The governor, in his speech to the Agricultural Society in 1842, referred to produce that was on display. The society's 1849 report indicated that an exhibition of produce took place again in 1848.[50]

Most of the settlement and agricultural activity was taking place in the eastern part of Newfoundland but there was some activity at outports in the western part of the colony. Thus Cormack, who crossed the island on foot in 1822, noted that at St. George's Bay, on the west coast, they all cultivated potatoes, and some kept a few cows. At Barachois River he reported that "some of the residents have well stocked farms, the soil being good. Oats, barley, potatoes, hay etc. are in perfection, and even wheat."[51]

Of one woman he wrote: "The cellar was full of potatoes and other vegetables for winter use. She was also an experimental farmer, and exhibited eight different kinds of potatoes, all possessing different qualities to recommend them." This woman had poultry, six milk cows, and other cattle.[52]

In 1827, the recording forms for the official statistics first made provision for agricultural entries but no entries were made. A notation stated: "It is impossible in the present state of the country to furnish any Account that can be relied on as accurate."[53] This notation continued in the annual statistics until 1836 although in 1834 there was the following addendum: "It may however be remarked generally that Agriculture is gradually increasing."[54]

A census taken in 1836 provided the first detailed recording of agricultural information and such information has been recorded more or less regularly since that date. Table I provides a summary for several of the years in the period under consideration in this study. Some comparable data for 1966 are included. These will be discussed later.

Table I
Population and agricultural statistics for a number of years[55]

		1836	1845	1857	1869	1966
Population	No.	73,705	96,506	122,638	146,536	493,396
Improved land	Acres	11,062	29,654	42,616	41,715	20,566
Hay	Tons	6,935	9,886	16,296	20,458	--
Wheat/barley	Bu.	--	---	1,932	747	--
Oats	"	10,310	11,695	9,038	11,150	--
Potatoes	"	311,832	341,160	228,572	308,357	--
Turnips	Bbl.	--	---	5,133	17,100	--
Other roots	"	--	---	1,801	8,857	--
Cattle	No.	5,835	8,135	19,886	13,721	8,554
Horses	"	1,551	2,409	3,509	3,764	1,061
Sheep	"	3,103	5,750	10,737	23,044	14,381
Goats	"	--	5,791	17,551	6,417	--
Swine	"	--	--	--	19,081	7,307

The data for potatoes in 1836 were recorded in barrels and converted to bushels at eight to the barrel.

The 1857 data for goats include swine. In the following census they were recorded separately.

The data in Table I indicate that there was a considerable increase in improved land from 1836 to 1857 and then it appeared to stabilize. To a large extent the same was true for the livestock population. The tabular data indicate that grain crops, other than oats, were of no real importance and even oats were a relatively minor crop. No yield data, on a per acre basis, were recorded but it is quite evident that, even with relatively low yields, the acreages in oats would be quite small in relation to the total improved area.

It is interesting that it was not until 1869 that swine were recorded separately and yet various references to production have indicated that swine were raised from the very earliest days of settlement. Judging from the 1869 data, compared to the combined data for goats and swine in 1857, it seems safe to assume that there was a substantial swine population in that year. Indirect support for this and for earlier swine production is in the small number of pigs imported, as seen in Table II. One might assume that these were primarily breeding animals and that local production filled most of the demand for fresh pork. However, it must be noted that both processed beef and pork were imported regularly.

That the livestock population was inadequate to meet the needs of the population is shown by the imports made each year.

Table II
Annual imports of live animals for selected years [56]

	Cattle	Sheep	Hogs	Horses
1827	1505	1987	--	33
1837	1816	1918	17	50
1845	2574	3290	63	117
1850	3042	3685	52	82
1857	3402	3741	120	114
1862	2014	4359	131	75

Comparison with the livestock populations recorded in the census, statistics suggest that a large proportion of the cattle and sheep imported were for slaughter and not for herd increases.

In the schedule of statistics to be recorded at census time there was a heading, "Improvements in Agriculture". Not until 1864 was there a notation under this heading and in that year we find the following: "Nothing of importance under this head. Attempts are being made to increase the stock of manure by the manufacture of guano from fish bones & offal etc., also by the more general use of lime and to improve the breed and increase the rearing of sheep, and to encourage the raising of flax."[55] This would seem to indicate that agriculture continued to be very much a secondary industry that was having a hard time making any real progress.

It has been noted that some of the governors took action to encourage agriculture, but direct support by the government appears to have been limited. One early step, directed in part at agriculture, was the erection of a Colonial House and a Market House, the latter to have stalls for the sale of produce.

Legislation Enacted

Various acts were passed and financial sums voted for the support of agriculture.

In 1839 there was an act to encourage the killing of wolves with a reward of five pounds for each wolf.[58]

In 1844 an act made provision for the sale of ungranted and unoccupied crown lands, the land to be offered at public auction with an upset price of two shillings per acre and a maximum of 100 acres per individual purchaser. The act required that the purchaser condition five percent of the land for cultivation within five years of purchase. This latter requirement was repealed the following year.

In 1845 financial support was granted for the Agricultural Society that had been formed in 1842. The estimates for 1845 contained: "The sum of two hundred pounds in aid of the funds of the Agricultural Society." This was repeated in 1846, reduced to £100 for 1847, 1848 and 1849, but then increased in 1850 to £150. Apparently there arose some dissension for in 1853 a change was made and the item now stipulated that: "The sum of One Hundred and Fifty Pounds towards the support Of the Agricultural Society: Provided that the sum of One Hundred Pounds, part of said sum, shall be expended for that purpose in the Outports of this Island."

The same terms and conditions were repeated in 1854 and 1855 and then the Assembly saw fit to be even more specific on how the money was to be spent as the 1856 estimates stated: "The sum of two hundred and fifty pounds toward the support of the Agricultural Society of Saint John's; to be expended as follows: The sum of one hundred and fifty pounds in the purchase of Seed, and of Cattle to improve the breed, in and for such Outport Electoral Districts as may require the same; and the remaining sum of one hundred pounds to be expended for the like purpose in Saint John's." This grant was repeated in 1857, 1858, 1859, and 1860, and then in 1861 was limited to £100 without any stipulation on how it was to be spent.

In 1862 an interesting item was added: "The sum of two hundred pounds toward the support of the Agricultural Society of St. John's, to be expended as follows: The sum of one hundred pounds in the purchase of Seeds, and of Cattle to improve the breed, in and for such Outport Electoral Districts as may require the same; and the sum of eighty pounds to be expended for the like purpose in St. John's; and twenty pounds for the introduction of Nova Scotia Hares into this Colony."

In 1864 a change was made to two separate grants, one for £100 for Conception Bay and one for £250 for St. John's and outports other than Conception Bay. This was continued in 1865, but the amounts were translated into dollars so there were $1,153.84 for St. John's and $453.54 for Conception Bay.

Three other acts are of interest. In 1851 there was: "An Act for preventing damage by unseasonable Burning and Firing of the Woods in the clearing of Land, and Burning of the Woods and Forests of this Island."

Then in 1861 dogs apparently had become a serious problem for there was passed: "An Act to Provide against the Destruction of Sheep and Cattle and Injuries to the Person, By Dogs." While the title of the act mentioned only sheep and cattle the body of the act referred to sheep, lambs, horses, goats, and cattle.

Finally, in 1864, it was deemed advisable to standardize weights for produce in: "An Act for Establishing the Standard Weight of Grain and Pulse, and to regulate the sale of Bread, Coals, and other Articles.

- Wheat shall weigh Sixty pounds
- Rye shall weigh Fifty-six pounds
- Indian Corn shall weigh Fifty-seven pounds
- Barley shall weigh Forty-eight pounds
- Oats shall weigh Thirty-eight pounds
- Pease shall weigh Sixty pounds
- Beans shall weigh Sixty pounds
- All such Grain shall be of dry, sound quality and all such weights shall be Avoirdupois.
- Potatoes, Turnips, Carrots, Parsnips, Beets and Onions shall weigh Sixty pounds
- Flax seed shall weigh Fifty pounds
- Hemp seed shall weigh Forty-four pounds."

In closing out this account of early agriculture in Newfoundland it is of interest to note that, contrary to the situation that will be discussed in the other provinces, no reference has been made to pests or diseases, other than to the unidentified potato disease. This disease seems to have made its appearance in Newfoundland at about the same time that a potato disease (possibly the same one) appeared in the potato crops of the other Atlantic provinces and caused numerous partial and complete crop failures. Lack of reference to insects and diseases in the available records suggest that these pests (other than the potato disease) were not important factors in Newfoundland during the period under review.

The agricultural industry in Newfoundland has not attained major status as indicated by the data for 1966 in Table I. The census provided data on land and livestock, but not on crop production. The *Canada Yearbook* provided production data for all provinces except Newfoundland, suggesting that these were minimal. Agricultural development in

Newfoundland has lagged behind that of other provinces partly because of early government policies, partly because of the limited amount of top quality land, and partly because of delay in developing the necessary infrastructure. For example, significant development of agriculture on the good lands of western Newfoundland did not occur until after the construction of the trans-island railroad in the last decade of the 19th century, which was later than the completion of the Canadian Pacific railroad across mainland Canada and later than the major agricultural development in the prairie provinces.

In looking back it appears that the early commentators who saw no real future for agriculture in the province were more nearly correct than those who considered that agriculture could develop to become an important element in the economy of the island.

Chapter II
Nova Scotia

Most historians date the beginning of agriculture in Nova Scotia from the establishment of Port Royal in 1605. However, the first steps toward agricultural development antedate this by almost 100 years. There were two earlier attempts at settlement and even 1605 does not mark the beginning of continuous agricultural activity.

The first effort at establishing a settlement was made by a Portuguese, Baron de Leury, who, in 1518, set out to settle Sable Island, a most unlikely site for a settlement. It ended in complete failure but some hogs and cattle were left there.[1] The next attempt was again on Sable Island, in 1598, under de La Roche. This effort has been given little attention, possibly because it ended in failure, but it is part of the history of the early development. Trudel has made a careful review of the evidence regarding this settlement, which lasted for five years. The first group of settlers consisted of up to 200 men and 50 women, mainly vagabonds and beggars, and they were settled on the north shore of the island where a storehouse and dwellings were built. The next year additional settlers and supplies were delivered. Contact was maintained annually until 1602 in which year no contact was made. In 1603, contact was renewed but by this time strife and desertions had reduced the number of colonists to 11 and when they were removed, another attempt to settle the island was ended. Little is known of the extent of agricultural activity but Trudel states: "The deportees benefitted from the game and fish they caught, as well as the numerous cattle unloaded previously by a Portuguese expedition, while their 'French gardens' supplied them with sufficient vegetables."[2] This would indicate that a considerable area was being cultivated to supply vegetables to the number of inhabitants on the island in the early years of the settlement.

We come now to the year 1605 when the first settlement on the mainland of Nova Scotia took place. In 1604 De Monts, with a group of settlers from France, settled on an island in the St. Croix river, which now forms the boundary between New Brunswick and Maine. Here they assembled buildings and took the first steps in agriculture. The first winter at St. Croix indicated clearly that this was not a desirable site and thus, in 1605, the settlement was dismantled and moved across the Bay of Fundy to Port Royal. Poutrincourt, who had come out with de Monts, had been given a grant of land which became the locale of the new settlement. In the move to Port Royal there apparently was not time for gardening as Lescarbots states: "And as for tillage of ground, I

The Trail Blazers of Canadian Agriculture

believe they had no fit time to do it, for the said Monsieur de Monts was not a man to be long at rest, nor to leave his men idle, if there were any means for it."[3]

Poutrincourt returned to France in late 1605 to obtain new supplies and settlers and he returned to Port Royal with them in 1606. There is conflicting information about whether livestock were brought out on this voyage. One report indicates that cattle were on board but that most, if not all, died *en route*. This same report notes that there were hogs that multiplied. Lescarbot, who came out with Poutrincourt that year, in writing of his experiences during his stay, states: "that I never made so much bodily work for the pleasure that I did take in dressing and tilling my gardens, to enclose and hedge them against the gluttony of the hogs."[4] At another point he states: "Yet it is well to say here, that our French domestic animals prosper very well in those parts. We had hogs which multiplied abundantly; and although they had a stye they lay abroad, even in the snow and during frost. We had but one sheep, which enjoyed the best possible health, although he was not shut up at night, but was in the midst of our yard in the winter time. M. de Poutrincourt had him twice shorn, and the wool the second year was reckoned in France two sous a pound better than that of the first. We had no other domestic animals save hens and pigeons, which failed not to yield the accustomed tribute, and to multiply abundantly."[5]

These statements by an observer on the scene would seem to rule out the presence of cattle, an assumption strengthened by a statement by Lescarbot in referring to the death of one of the settlers: "And, if we had had half a dozen kine, I believe that nobody had died there."[6]

Poutrincourt brought with him his cousin, Louis Hebert, who was to have an influence on the agriculture of Port Royal. His time there whetted his appetite for Canada and later led him to Quebec where he was to become the father of agriculture in that province. Hebert was an apothecary by profession but had a vital interest in agriculture. His was the same profession as that of another man who was to have a great influence on Canadian agriculture in later years, namely, the founder of the Experimental Farms System, W. Saunders.

Poutrincourt arrived at Port Royal on July 27, 1606. As an indication of the energy with which he advanced the work at the settlement we read: "The Friday, next day after our arrival, Monsieur de Poutrincourt affected to this enterprise, as for himself, put part of his people to work in the tillage and manuring of the ground, whilst the others were employed in making clean the chambers, and everyone to make ready that which belonged to his trade."[7] There is no record of any crop having been seeded prior to his arrival and no specific information that a crop was harvested that year. It is recorded that: "in fifteen days after his arrival thither, he sowed it with French corn, as well wheat and rye as with hemp, flax, turnipseed, radish, cabbages, and other seeds, and the eighth day following he saw that his labour had not been in vain, but rather a fair hope, by the production that the ground already made of the seeds which she had received."[8]

Later that year the following was reported: "The public rejoicing being finished Monsieur de Poutrincourt had a care to see his corn, the greatest part whereof he had sowed two leagues off from our fort; up the river l'Equille, and the other part about our said fort; and found that which was first sown very forward, but not the last that had been sowed the sixth and the tenth days of November, which notwithstanding did grow under the snow during the winter, as I have noted in my sowings."[9]

It appeared that success was in the making and the spring of 1607 arrived full of hope and plans for the future. But the hopes were not to be fulfilled. De Monts had lost his charter and it became necessary to abandon the settlement. Referring to the boats leaving, Lescarbot wrote: "And on the 30th of July the other two went away. I was in the great one, conducted by Monsieur de Champdore. But Monsieur de Poutrincourt, desirous to see an end of our sweet corn tarried till it was ripe, and remained there eleven days yet after us."[10]

It is evident that agricultural activity did not begin until 1606 and was not continuous. There is even some doubt that it was continuous after the return of Poutrincourt in 1610 with new resources. On this occasion he brought not only settlers and supplies, but also cattle.

On his arrival he found the buildings and grounds as they were when he left in 1607. He divided the land among the new male settlers who were to be tenants, paying a small annual rent. Work was undertaken immediately to clear more land and prepare it for sowing. Rameau writes: "Il avait amené avec lui des bestiaux, et reprenant les charrues on se mit immediatement au labour, afin de preparer les semailles d'automne."[11] This is the first record of the plow being used and there are some who doubt the correctness of the statement. It is feasible if we assume that some of the cattle brought over were draft animals.

Some support for the statement is the record of Lescarbot, in writing of Argall's raid in 1613: "Whence they crossed to Port Royal, but found no one there for M. de Biencourt suspecting no enemy had put to sea, and part of his men were engaged in ploughing two leagues from the fort."[12]

Supplies and additional settlers were sent in the years following 1607 and some time during the years 1610-1612 horses appeared on the scene. All went well until late in 1613 when a fleet from New England, under Argall, raided the settlement. Parkman gives a somewhat colourful description of what happened. "The magazines were by no means empty, and there were cattle, horses, and hogs in adjacent fields and enclosures. Exulting at their good fortune, Argall's men butchered or carried off the animals, ransacked the buildings, plundered them even to the locks and bolts of the doors; then laid the whole in ashes... Port Royal demolished, the marauders went in boats up the river to the fields where the reapers were at work. These fled, and took refuge behind the ridge of a hill whence they gazed helplessly on the destruction of their harvest."[13]

This was the end of Poutrincourt's efforts to establish a settlement. Port Royal was abandoned. Some of the settlers remained in Acadia to continue as fur traders and it seems that a small amount of agricultural activity continued, not only at Port Royal but at other trading posts. For example, Denys, in reporting on a visit at La Tours fort at Cape Sable in 1635, wrote: "During the time I was there a Recollet father arrived, to whom the wife confided the pleasure she had in seeing me. Then I discoursed with the Recollet who gave me an account of his garden, he invited me to go see it, and I accepted... We arrived at the garden, (and) he told me that he had cleared it all alone. He might have had about a half arpent of ground, and he had there a quantity of very fine well-headed cabbages, and of all other sorts of pot herbs and vegetables. He had also some apple and pear trees, which were well started and very fine, but not yet in condition to bear, since they were brought small from France, and had been planted only the preceding year. I was much pleased to see all this, but more when he showed me his peas and his wheat, which he had sowed. The young La Tour had also a garden near his fort, with wheat and his peas which were not so carefully cared for as were those of the Recollet."[14] This is the first specific reference to apple and pear trees in Nova Scotia.

With the withdrawal of Poutrincourt there was an interlude of 15 years when agriculture played, at best, a very minor role. It was not until Alexander, with his Scottish settlers, arrived at Port Royal that real agricultural activity was renewed. There is some difference of opinion as to the actual date of his first arrival. Lanctot[15] sets the date as 1627 whereas Coleman[16] indicates 1629. There is agreement that the effort was a substantial one, with four ships and 70 settlers. In 1629, Alexander also established a settlement at Balines, on the coast of Cape Breton, but this was destroyed the same year, by the French. The settlement at Port Royal was resupplied and maintained until 1632 when the area was returned to French sovereignty and the Scots had to leave.

One could choose the arrival of the Scots as marking the point when continuous agriculture began, but it would be almost as correct to choose 1632, the year that the French repossessed the colony. In this latter year the main settlement was not at Port Royal. The new settlement was under the auspices of the Company of One Hundred Associates, who sent out Chevalier de Razily with settlers and supplies.[17] He chose Le Have as the site for settlement as it was thought to be a better location for trading purposes and trading with Indians still was the main source of financing settlement. This was a large-scale venture, consisting of over 300 men and women and all sorts of supplies and equipment, as well as a variety of farm animals. Land clearing and development of needed facilities took place during the next three years, but then it was decided to move the main part of the settlement to Port Royal and retain Le Have mainly as a trading post. According to Lanctot: "the land was blacker, richer, and already cleared. By 1641 the farming concessions there covered more than a square mile and with the years spread far beyond and above the post. Wheat production rose rapidly, and other cereals were harvested in increasing quantity, and the herds of cattle and swine multiplied."[18] It can be said that agriculture finally had a solid foothold in what is now Nova Scotia.

Many of the settlers at Port Royal were from the settlement at Le Have but additional settlers were brought directly from France. The occupations of those selected to come are listed by Lanctot: "The list of emigrants who took their places aboard the Saint-Jean at La Rochelle on April 1, 1636, reveals the scope of the company's plans: eight ploughmen and one lumberman from Burgundy; nine ploughmen, one farrier and swordsmith and his valet, one cooper, two tailors, one cobbler, a vine grower, an armourer and a flour miller, all from Anjou; three mill carpenters from the Basque country; five salt makers and three sailors from Aunis. Eight women with nine children also came aboard."[19]

Marshland Development

Some of the settlers who came after 1632 were from a part of France where dyking and draining of tidelands was practised. They applied these methods to the marshes at Port Royal and this was the beginning of what was to become a very important aspect of the development of many parts of Nova Scotia and New Brunswick. In fact, the relative ease of marsh development, as compared to clearing upland forests, was mainly responsible for the next stage of development in the province.

That the marshland development was not an unmixed blessing is suggested by the report of Gargas, who spent the winter of 1687-88 in Acadia. His evaluation of the situation at that time is of considerable interest. "It would also be a good plan to force the inhabitants to clear the higher ground. Most of them, as at Port Royal, Mines, etc., take pleasure in building levees in the marshes where they sow their wheat. This (system) does them harm for several reasons, the first of which is that the grain which they sow in their marshes is very small in grain and yields half bran and also does not keep. The tides often burst their dykes and flood their fields, which are not productive again for several years after being soaked with salt water, and the repairs which they have to make each year cost them much money and effort.... (Moreover) they run the risk of a high tide coming just when they are ready to harvest their crops and destroying all their hopes, which is dangerous. Again because they plough their marshes, they are short of fodder, which means that they are unable to feed their cattle, either those useful

The Trail Blazers of Canadian Agriculture

for labour or those needed for food for the inhabitants. Whereas, if they would clear the higher land, they would have only the initial difficulty, they would be safe from all accident, they would have very good grain, and they would have enough marsh land to support an enormous number of cattle; this would be of certain use to them for their own needs and for the provisioning of ships which might come ashore there. The inhabitants appreciate the force of these arguments, but, having already built the dykes, they do not wish to undertake new labours, and the country will always remain unchanged, especially if the Governor allows the young sons of the colonists to go and settle elsewhere along the coast, where they do nothing but hunt and negotiate with the natives."[20]

Port Royal was the main settlement for many years and the only one where agriculture played a major role as the primary source of livelihood for the inhabitants. It was from here that the main thrust of further development took place. However, the picture of early development is not complete without noting the activities of Nicolas Denys. His is in many respects a continuing hard-luck story as he was in almost constant conflict with competitors. He was a man of considerable enterprise and drive. His basic income was derived from fishing, but at times he also was involved in lumbering and trading with the Indians. However, our primary interest is in his agricultural activities in which he pioneered at numerous locations in Nova Scotia and New Brunswick. It seems that at all of the sites where he established posts he immediately undertook farming activities.

One of his early posts (1653) was at Saint Peters on Cape Breton and of this he wrote: "I have a good eighty arpents of cultivated land, which I have sown every year before my fire."[21] The fire in 1668 destroyed his holdings and forced him to give up at this site. Of this he wrote: "Nothing was saved except half a cask of brandy, as much wine, with about five hundred sheaves of wheat, which we had much trouble rescuing from a barn where the fire had not yet caught."[22] From another notation it is evident that he was not alone in agricultural activity at this site: "A person named Montague who had been with me, and whom I had married to one of the servants of my wife, worked upon his own account at Saint Pierre on the Island of Cap Breton. He had there some six arpents of good land cleared and without stumps, where he harvested good wheat, peas and beans, and this by means of the advances I had made him."[23]

Another of his posts was at Chedabucto (established about 1660) of which he wrote: "The grass there is fine, and grows as high as a man. It was the fodder for our cows when we were at Chedabucto, which is two leagues further into the head of the bay. This was the place which I had chosen for constructing my storehouses in order to establish a sedentary fishery. I had a hundred and twenty men at work there, as well at building as at farming. I had about thirty arpents of land cleared, of which part was in crops. All these lands are returned to their primitive state and the buildings are ruined."[24]

Despite setbacks of various kinds the settlement at Port Royal continued to grow, even during the occupation by the English, which occurred in 1654. Shortly after the French regained possession in 1670, the first census was taken in 1671 showing that the human population included 68 families consisting of 63 couples, 5 widows, and 227 children. They owned 829 head of cattle, 399 sheep, and had 417 arpents of land under cultivation.[25] For some reason the census contains no record of either horses or pigs though other records indicate that both classes of stock were being raised at the settlement.

By 1671 certain pressures had developed that caused the first move toward hiving off from the parent colony at Port Royal. One of the factors sometimes cited was the shortage of marsh land at Port Royal, but with only about 400 arpents under cultivation it may be questioned whether this was important. The settlers at Port Royal were not unfamiliar with other areas around the Bay of Fundy which they had been visiting from time to time in hunting, trading, and fishing expeditions. It was primarily young people who ventured into these new areas but it is interesting that one of the first to leave the old colony was an older man, Jacob Bourgeois, who had come to the colony originally as Aulnay's surgeon. His first move was made in 1672 or 1673 when he went with other members of his family and settled at Chignecto.[26] It was not long before a number of others were settled in the same general area.

Not only did the new area attract settlers from Port Royal but Michel Leneuf de la Valliere, from Quebec, was granted a seigneurie of about 1000 square miles in the area in 1678, but with the stipulation that the prior settlers were to be excluded from the grant. There was gradual development in the area and when Demeulles, an official from Quebec, visited there in 1686 he wrote: "Around it are extensive meadows (marshes) which are capable of feeding one hundred thousand head of horned cattle, its grass being termed 'misotte', very suitable for fattening all kinds of animals."

"There have already been erected there more than twenty-two dwellings upon the small eminences which the inhabitants have chosen in order to have access to the meadows and the woods. There is not a single one of its inhabitants who has not three or four buildings quite suitable for the country. Most of them have already twelve or fifteen horned cattle, and even as many as twenty, ten or twelve pigs and as many sheep."[27]

The report went on to state that the region had 22 families, of which only 19 lived there continually, numbering 129 men, women and children. The other three families spent part of their time at Port Royal. In the entire region there were 270 cattle, 118 sheep, 189 pigs and 576 arpents of cultivated land. These figures are at slight variance with the census data which show the following for that year and two later years.

Table I
Census data for the Beaubassin area for three separate years[28]

	1686	1693	1707
Human population	127	119	226
Arpents cultivated	426*	157	286
Cattle	236	309	510
Sheep	111	280	500
Swine	189	146	328

*Considering the data for the later years it is likely that this was not all cultivated.

Once the movement from Port Royal got under way it was not long before settlements also were started at Minas and Cobequid. These new settlements also prospered and grew rapidly as can be seen from the following census data for 1707.

Table II
Basic data from the 1707 census[29]

	People	Arpents cultivated	Cattle	Sheep	Swine
Minas	498	291	766	718	639
Port Royal	458	392	963	1245	974
Cobequid	67	101	180	128	114
Beaubassin	226	286	510	500	328

When the English took permanent possession in 1713 the settlements at Beaubassin and Minas had populations as great as, or greater, than Port Royal. It is estimated that at that time the total population was in the range of 2000. Some estimates place the number considerably higher but, in view of the 1707 data, this seems to be unlikely.

Following the occupation of Acadia by the English there was relatively little change in the pattern of life and agricultural development until the middle of the 18th century. There was increasing trade with the French population at Louisbourg, and even with the New Englanders, much to the distrust of the English at Port Royal, now renamed Annapolis Royal. An example is found in the letter from John Doucett to the Lords of Trade and Plantations, written November 15, 1718: "That vessels from Cape Britton, Spring and Fall come to Minis which is about twenty leagues higher in the Bay of Fundy than we are, and the Greatest Settlement for Growth of Corn att Present in this Colony. The French from Cape Britton, bring Wine Brandy & Linnings which they can afford Four pence and Six-pence in a Yard Cheaper than our Traders can Possibly doe. And take from thence nothing but Wheat & Cattle which they kill there & salt up, and from Chignecto which is twenty leagues

Higher in the Bay than Minis. They drive Cattle over to Bay Vert and from thence Transport them, which is not only a great Detriment to our Traders who cant Sell their Goods but will raise the Price of Provisions & impoverish the Collony, or att least make it of more benefit to France, then to us if not hindered, who likewise Carry all the Small Furs they can out of the Country...There being at this time two sloops sailed from Minis with Severall Hundred Bushells of Wheat & severall Head of Cattle to Cape Britton, the owners of which were So Insolent to tell our Traders that came in there with their Cargoes that they had nothing to do there and they would be there again in the Spring for more wheat, which is so true that the Inhabitants of Minis are Dayly thrashing their Corn to get their Loading ready at their return..."[30]

Much later, in 1743, this problem still existed, as we see in a report by Herbert Newton: "They in a clandestine manner supply the French at Lewisburg and St. Johns with 6 or 700 Head of cattle, and about 2000 sheep in a year in short Lewisburg in my opinion would starve if it was not for them, tho' at the same time they do this Our garrison at Annapolis Royal and Canso, which is in their Neighbourhood, are in great want and can get neither Beef nor Mutton but at Great Expenses from New England, and Acadians before mentioned have their woollen and Linnens and most of the necessary they want from Lewisburg, and are in a manner dependent on the French tho' they live in Nova Scotia."[31]

The basic problem under English rule was that in the early years no non-French settlers came to Acadia so that the population continued to be predominantly Acadian. Governor Phillips, some years after the English occupation made a plea to the Lords of Trade and Plantations for new immigrants. "In Obedience, therefore, to what your Lordships require of me, I answer (in respect to Nova Scotia only) that it is my humble opinion, that the chief encouragement wanting toward the soils Cultivating, and improvement thereof, is the creating of Two or Three Forts in proper places, with an addition of 2 or 300 Men, to Garrison such Forts. This may invite a new Set of People that are Protestants, to venture their Lives, and Fortunes, under the Protection of that Government, for as to the present inhabitants, they are rather a Pest, by Incumbrance, than an advantage to the Country, being a proud Lazy, obstinate, and untractable People, unskillful in the method of Agriculture, nor will be led or drove into a better way of thinking, and (what is still worse) greatly disaffected to the Government. They raise, (tis true) both Corn & Cattle on the Marsh Lands, that wants no clearing, but they have not in almost a Century, cleared the Quantity of 300 Acres of Wood Land. From their Corn & Cattle, they have plenty of Dung for Manure, which they make no use of, but when it increases so as to become troublesome, then instead of laying it on their Lands, they get rid of it, by removing their Barns to another Spot."[32]

No doubt this was a somewhat biased evaluation of the inhabitants and their practices, though not completely at variance with observations made by other less biased individuals as we shall see in a later section in which the agricultural practices are reviewed.

During the early years of English possession of Acadia practically no English immigrants arrived in the colony to take up land and engage in agriculture. It was not until Halifax was established in 1749 that the first major influx of new people arrived. The first immigrants were not primarily agricultural people but rather artisans and workmen for building fortifications and housing. It is interesting to note the evaluation of this first lot by Governor Cornwallis: "The number of settlers Men and Women & Children is 1400 but I beg leave to observe to your Lordships that amongst these the number of industrious active men proper to undertake and carry on a New Settlement is very small — of soldiers there is only 100 — of Tradesmen Sailors and others able and willing to work not above 200 more — the rest are poor idle worthless vagabonds that embraced the opportunity to get provisions for one year without labour, or Sailors that only wanted passage to New England...There are among the Settlers a few Swiss who are regular honest industrious Men, easily governed and work heartily. I hope your Lordships will think of a Method of encouraging numbers of them to come."[33] Apparently this was accomplished for in 1753 the Lunenburg settlement of German and Swiss immigrants was begun. But more of this later.

Subsidies Instituted

The need for increase in agricultural production to feed the increased population was recognized by the governor and his council and in 1752 the first agricultural subsidies in Atlantic Canada were instituted to encourage production. The details of this are of sufficient importance to warrant citing. The *Halifax Gazette* of April 13, 1752 carried the following public notice dated April 8:

"It is by His Excellency the Governor, and his Majesty's said Council, Resolved, and by the Authority of the same is Enacted, That, For the Encouragment of all Persons who may be disposed to exercise their Industry in Husbandry and Agriculture within this Province, the following Bounties shall be allowed and paid out of the Treasury of the Province, viz."

"Upon all land which has been granted by his Excellency Governor Cornwallis, or that shall hereafter be granted by him or his Majesty's Governor or Commander in Chief of said Province for the time being, which land shall not have been cleared before the said Grant thereof, and which shall within Twelve Months from the date hereof, be fenced with a substantial Fence, not less than four feet high, and be cleared of all the Underwood and Brush, and shall have the Trees thereon felled, (excepting a Number not exceeding Ten on each Acre), and shall be sowed either with English Hay-seed, or with any Kind of English Grass, or with Hemp or Flax-Seed, the Sum of Twenty Shillings per Acre, for each Acre so bro't to and improved, within the said Term of Twelve Months from the Date hereof."

"The said Bounty to be paid to the respective Owners of the said Lands, their Heirs, Executors, Administrators, Assigns, or certain Attornies, upon a certificate of the before mentioned Improvements having been made, being produced to the Treasurer of this Province, under the Hand of such Person or Persons as shall be, by his Excellency the Governor appointed to inspect and certify the same."

"Also there shall be allowed and paid (in like Manner) the Sum of Two Shillings, per Hundred Weight, upon every Hundred Weight of English hay which shall within Eighteen Months from the date hereof, be produced from any of the before-mentioned Lands; and Two Shillings per Bushel upon every Bushel of Wheat, Barley, or Rye; and One Shilling per Bushel of Oats produced therefrom within the said Term of Eighteen Months."

"Also there shall be allowed and paid (in like Manner) the Sum of Three Pence per Pound upon every Pound Weight of merchantable Hemp, which shall be bright, well cured, and water rotted, of four Feet at least in Length, and cleansed for Use."

"Also Three Pence per pound upon every Pound Weight of Merchantable Flax, either dew or water rotted, well cleansed, and fitted for the Market; Which shall be produced from any of the aforesaid Lands, within the Term of Two Years from the Date hereof."[34]

This was a portent of the future, the first real expression of government concern for agriculture and the first indication of a policy of subsidizing agriculture that has continued in a variety of forms to the present time.

Somewhat related was the assistance provided to new immigrants in the form of equipment, supplies, and livestock. The earliest example of this was in the settlement of Halifax itself but was extended to the settlement of Lunenburg and became of major importance in the later settlement of the Loyalists.

The Lunenburg settlement was undertaken in 1753 when about 1600 Swiss and German immigrants were settled there. The basis for this is outlined in a letter from Governor Hopson: "I propose to send out the Foreigners in three days; they go to Merleguish, a harbour about 16 leagues to the westward from this place, where there has been formerly a French settlement, by which measure there is between 3 and 4 hundred acres of cleared land which is to be equally divided amongst the settlers who consist of 1600 persons."[35]

The first contingent arrived at Lunenburg on June 8 and the second on June 17. It was reported that: "it is fine open Country, the Soil exceeding good, the Grass almost as high as a Man's knees, the Fruit Trees all in Bloom, etc.."[36] Despite some bad weather Col. Lawrence could report on July 11: "Most of them are well under cover. All of them have gardens, & many of them good Framed Houses. They have cut on ye whole a Considerable quantity of hay."[37]

In a report on July 2 we find the first specific reference to potatoes in Nova Scotia: "The Warren Capt. Adams came in at ye same time with the Trivelt having on board only about 12 or 13 bushels of Potatoes. I shall take them out & distribute them in ye best manner I can, but I fear not much to ye satisfaction of ye people as ye quantity is so small."[38]

An interesting advertisement appeared in the June 22, 1754 issue of the *Halifax Gazette*: "The Government of this Province having occasion to contract for the following Quantities of Cattle, to be delivered in Health and good Condition, at the Risque of the Contractor, at Lunenburg, for the Use of the Settlers there:

  80 Cows, a Calf to each
130 Breeding Sows
450 Yews, a Lamb to each
200 Goats
   3 Bulls
   2 Boars
   2 He Goats
   4 Rams "[39]

The year 1775 saw the expulsion of the Acadians which left the developed land unoccupied. It was not until 1760 that the first real influx of new settlers occurred with the arrival of New Englanders, followed by settlers from Britain. Although this movement was relatively short-lived, much of the land previously occupied by the Acadians was brought back into production. The final major immigration was by the Loyalists, beginning in 1783.

It is timely now to review the agricultural practices that had been developed and used during the 150 years that Acadia had been occupied by Europeans and the status of the agricultural industry at the end of this period. Descriptions of the agricultural practices of the French period are sketchy and do not provide detail on many important aspects. Through most of the French regime agriculture was quite primitive. The first reference to the use of the plow was in 1610, but there is reason to believe that this implement did not come into widespread use until much later and the spade, pickaxe, and hoe continued to be important tools for cultivation of the land. For most of the time prior to the English occupation, agriculture was essentially a subsistence industry as there was no strong incentive for production beyond local needs. In fact, for much of this time frequent raids by New Englanders and pirates served to discourage major development.

The need of the settlers for manufactured goods was limited, but such needs were met by trade with France, and even with the New Englanders, but was based on fur rather than on agricultural products as the medium of exchange. After the English occupation the Acadians continued to trade in agricultural products, primarily with the French at Louisburg and, in return, French goods were received from that source.

A composite picture of agricultural practices and crop production can be obtained from the writings of several observers. A good description of the agricultural situation in Acadia near the end of the 17th century is given by Governor Villebon: "It is more than 60 years since Port Royal was founded and the work of clearing the land and enclosing the marshes began. The latter have, up to the present time been very productive, yielding each year a quantity of grain such as corn, wheat, rye, peas and oats, not only for the maintenance of families living there but for sale and transportation to other parts of the country."

"Flax and hemp also grow extremely well, and some of the settlers of that region use only the linen, made by themselves, for domestic purposes. The wool of the sheep they raise is very good and the clothing worn by the majority of the men and women is made of it."

"Port Royal is a little Normandy for apples. They might have a great many more, and could easily make cider; but apart from the fact that they are not very industrious and most of them only work for a bare living, they neglect the propagation of fruit trees for use as the country opens up. Calvilles, Rambours, Reinettes, and three or four more varieties of apples are found at Port Royal, and russet pears. There were other varieties of pears, and cherries, including the Bigarotiers, which they have allowed to go to ruin. There is an abundance of vegetables for food, cabbage, beets, onions, carrots, chives, shallots, turnips, parsnips, and all sorts of salads; they grow perfectly and are not expensive. Wheat, however, costs 40s a bushel, which, according to the test made by Sr. de Villebon in this country, weighs only 41 1/2 lbs. Before this war it was worth only 35s, and that by a ruling of the late M. d'Aulnay. Fine green peas are no more expensive, but they are the same rate as wheat."[40]

More detail on certain aspects is provided by Dierville, writing at the end of the 17th century. It may be noted that he differed with Gargas about the relative value of upland and marshland.

"It costs a great deal to prepare the lands which they wish to cultivate; those called Uplands, which must be cleared in the Forest, are not good, & the seed does not come up well in them; it makes no difference how much trouble is taken to bring them into condition with Manure, which is very scarce; there is almost no harvest, & sometimes they are abandoned. To grow Wheat, the Marshes which are inundated by the Sea at high Tide, must be drained; these are called Lowlands, & they are quite good, but what labour is needed to make them fit for cultivation! The ebb & flow of the Sea cannot easily be stopped, but the Acadians succeed in doing so by means of great Dykes, called Aboteaux, & it is done in this way; five or six rows of large logs are driven whole into the ground at the points where the tide enters the Marsh, & all the spaces between them are so carefully filled with well-pounded clay, that the water can no longer get through. In the centre of this construction, a Sluice is contrived in such a manner that the water on the Marshes flows out of its own accord, while that of the Sea is prevented from coming in. An undertaking of this nature, which can only be carried on at certain Seasons when the Tides do not rise so high, costs a great deal, & takes many days, but the abundant crop that is harvested in the second year, after the soil has been washed by Rain water, compensated for all the expense. As these Lands are owned by several Men, the work upon them is done in common; if they belonged to an Individual, he would have to pay the others, or give to the Men who had worked for him an equal number of days devoted to some other employment; that is the manner in which it is customary to adjust such matters among themselves."[41]

"With the exception of the Artichoke & Asparagus, they have an abundance of every kind of vegetable, & all are excellent. There are fields of White-headed Cabbage & Turnips, which are kept for the entire year. The Turnips are put in the cellars; they are tender & sweet, & much finer than in France; moreover they may be eaten cooked in the ashes, like Chestnuts. The Cabbages are left in the field after they have been pulled up, the head down & stalk in the air; in this way they are preserved by the snow which comes & covers them to a depth of five or six feet, & they are only taken out as needed; the Settlers never fail to put some in the cellars as well. Neither of these vegetables goes into the Pot without the other, & nourishing soups are made of them, with a large slice of Pork. It is necessary, above all else, to have a great many Cabbages, because the People eat only the hearts, & the Pigs are given what is left; during the winter it is their only food, & these gluttonous animals, of which there are a great number, are not satisfied with a small quantity."[42]

"The Sheep, likewise, are admirable, & as large as those of Beauvis; they are also reasonable in price, the finest, well fattened, being worth only eight francs; but, as they are kept for their wool, few are for sale. Like the Cattle, they are, ordinarily, fat only in the Autumn, because little grass grows on the Uplands, & there is no other place where they can graze. Cows are never killed; the Settlers are too fond of milk, & this, perhaps, is the reason they do not care for Veal, for, it is the peculiarity of the Cows in that Country, that if a Calf is taken from its Mother, her udder yields nothing more...Although butter is made in the Country it is not good, & each Settler keéps only a very small supply, preferring to use the milk."[43]

Morris, at mid-18th century, provides the last detailed picture of the cultural practices of the Acadians: "Their Tillage is performed with much ease being entirely free from Stones; that two yoke of Cattle is

Sufficient to Plow up their Stubble which is usually done in the Fall of the Year, it is Plowed up in Ridges about five feet Wide for the sake of Drainage of the Water into Trenches which are cutt in the Meddow and inclose about four or five Acres. These Trenches drain the Water into Channells which were formerly Creeks in the Meddow, and thus in the Ebb of the Tide all the fresh Water is drained into the Sea which without these Trenches and this manner of Plowing would rest upon the Land and render it unfit for Tillage. The Land thus Plowed lays open to the Winter Frosts and by that means the Gleb is so dissolved and Mouldred that in the Spring in the beginning of April they have no further trouble but to sow their seed and Harrow it in and from thence good Crops are produced from Year to Year... The upland is made use of for the Production of all sorts of Roots and Garden stuffs but very little is used by them for either Grass or Grain. Some particular Rich spotts do Produce good Wheat and other Grains, but that is not general, the Land is well adapted for English Grass, but as they have fodder enough from the Marshes without any other Labour than that of Mowing and Making they have hitherto much neglected the Improvements of their uplands which require a great deal of Labour to Clear, & which afterward must be fenced in, to prevent their Cattle destroying it who are turned loose into the Woods all the Summer Season."[44]

Somewhat later, after the Acadians had been expelled and a new group of settlers had come in, Deschamps provided a more sympathetic evaluation of the Acadians and their practices. He noted that, in conversation with some of the older Acadians, he had learned that they had tried a number of systems and that they had arrived at the most profitable on the basis of experience. He suggested that if the new settlers would follow the successful practices of the Acadians they would be more successful than by ignoring them. With reference to the dyked lands he noted: "The Acadians found these lands of more Easy Tillage, and more profitable, than the upland, and therefore did not clear much of the latter — yet the upland in General is very fertile & by proper husbandry will yield good Crops, not so large as the marshland."[45]

In 1782 Deschamps provided a sketch of the province which gives some indication of the developments that had taken place.

Table III[46]
Population and agricultural statistics
for certain areas in Nova Scotia in 1782

Locations	Inhabitants	Horses	Oxen	Cows	Young cattle	Sheep	Cleared land	Dyked land
Windsor	600	140	200	400	300	1000	2000	2000
Newport	900	300	100	400	600	700	1500	700
Falmouth	280	80	200	300	500	1200	2000	800
Cornwallis	1200	200	300	600	1000	500	4000	1000
Horton	1000	2000	3000	5000	--	--	2400	200

The Trail Blazers of Canadian Agriculture

The other districts of Cumberland, Annapolis Royal, Cobequit, and the sea coasts raise great quantities of Cattle for sale, and grain sufficient for their subsistence — the high price of labour in the country prevents the Inhabitants from being able to raise such quantities of Wheat and Rye as well as other grains as they could Otherwise, but they raise sufficient in general for the use of the Inhabitants — and a Scarcity having happened at Halifax the last Winter 1781 there was brought to it from Horton, Cornwallis and Windsor, upwards of 60 tons of wte flour — and great quantities of Hay, and Oats, and left the inhabitants a sufficiency — a peace would make great odds in this particular.

The Wheat weighs from 58 to 60 lb. per Bushel
    Rye   "   "   50 to 54 lb.
    Barley "   "   50
    Oats   "   "   35 to 42 lb.

The quantity of wheat sown on an acre of marsh land is 1 1/2 Bushels, it yields from 20 to 25 Bushels per acre.

    Barley Seed 2 Bushels and Yields 25 Bushels per acre.
    Oats   " 3   "   "   "   39 to 40   "   "
    Pease " 1   "   "   "   12 to 15   "   "
    Rye   " 2   "   "   "   25   "   "

Upland wheat 2 Bushels produce 16 to 20 bushels per acre
    Barley 2   "   "   20   "   "   "
    Oats   3   "   "   25 to 30   "   "   "
    Pease 2   "   "   12 to 15   "   "   "
    Rye   2   "   "   15 to 20   "   "   "

At about this same time Robinson and Rispin made a tour through Nova Scotia and wrote a comprehensive report from which the following is an extract: "When they break up the swarth land in the marshes, they plow it about the fall, and sow it in the spring with wheat, which grows very well. We saw fine wheat, growing upon the marshes, and as thick as it could stand. The soil is exceedingly good, and several yards deep. The French have grown wheat for fourteen or fifteen years together without a fallow, and the land brought good crops to the last."

"Their cattle are but small, much like our Lancashire beasts, but not quite so large. They are lively-looking cattle, with fine horns. They keep many oxen, with which they till their lands, and use them in all draughts. We have seen from one to four pair of oxen at one team, both at plow and at a wain, which they call carts, without any horse at them. They are in general good draught beasts and are as tractable and observe the driver's word, as well as our horses in England. They work their oxen until they are eight or ten years old before they feed them, and they in general grow to be good beasts. During our stay at Cornwallis, we saw a pair which had been fed, sold to a butcher at Halifax for thirty-three pounds fifteen shillings. They do not use whips in driving. We never saw any in the country, instead of which they make use of long rods. The French used to yoak their cattle by the horns; but in those parts they yoak them now after the English method."

"The horses are small, chiefly of the French breed, about fourteen hands and a half high, plain made, but good in nature. They seldom draw with any, so that few keep more than one or two for their own riding; they all naturally pace, and will travel a long way in a day. They are very dear; a horse that would sell for about six pounds in England, would fetch ten with them. Their method of breaking them is very extraordinary. They yoke a pair of oxen to a cart, and tie the horse to it, and drive away till they have rendered him quite gentle. They then put on the bridle, and he is mounted without more to do."

"Their cows, like the oxen, are but a smallish breed, and their management of them so bad, that they give but a small quantity of milk; for they fetch them up early every evening to milk, and let them fast till seven or eight o'clock the next morning. Mr. Robert Wilson, who went there from Helperby, nigh Boroughbridged, in the county of York, bought an estate at Granville. He let his cows out all night in their pastures, and the little time he had them when we were there, which was about three weeks, they gave near double quantity. A pretty good cow and calf will sell at Cumberland for about five pounds ten, or six pounds. It is very common amongst the wealthy farmers, to let out to the poorer sort of people their cows for twenty shillings a-year. There are some that have from ten or twenty let in this manner. They generally value the cow when they lend her out, and if any improvement is made, the borrower has a proper consideration; but if she be worse he must make a suitable satisfaction. They let out brood mares and sows after the same manner."

"Their method of rearing calves is somewhat singular; as soon as they go to milk, they turn out their calves which suck one side of the cows, as the women milk on the other, and when they have done they are put up again, and continued to be fed in this manner till they are three or four months old, when they are turned out to grass... "

"The sheep appear to be of the Spanish breed, are long legged, loose made, and have short but fine wool. They clip four, five, and some six pounds, which they sell for eighteen pence a pound."

"The pigs are a very different breed, much inferior to any we ever saw in England. They feed them very fat with Indian corn, pumpkins, or potatoes. They keep their pork and beef always in pickle, and never dry as is customary in England."

"The inhabitants are of different countries, though chiefly from New England, Ireland, and Scotland... Nothing can be said in favour of the inhabitants, as to the management in farming. They neither discover judgment or industry."[47]

Practices Criticized

A strong indictment of early 19th century agricultural practice was made by 'Agricola' (John Young),[48] who became noted for a series of letters on agriculture, the first of which appeared in the *Acadian Recorder* in 1818. In his criticism he emphasized the lack of knowledge displayed by most of the farmers about the basic principles of farming and their application. Manure was not used, cultivation was inadequate, weeds flourished, and rotation of crops was practically unknown. All in all he painted a picture of agriculture at a very low level of development. There were several underlying causes for this general state of affairs, a significant one being the limited market for the products.

Some steps had been taken to improve the situation and these were stimulated and increased by Young's letters. The most effective action for improvement came through the agricultural societies. The first one on record was organized in Halifax in 1786, but there is an indication that it did not survive.[49] However, in November 1789, a Society for the Promoting of Agriculture in the Province of Nova Scotia was established at Halifax under the patronage of Lieutenant Governor Parr. The purpose of the society, which had representatives from the various parts of the province, was outlined as follows:

"An Institution, which has, for its object, the real Welfare and Prosperity of the Province, cannot but meet with the most generous and liberal support, and those who have formed this Society freely invite communications upon all Subjects comprehended within their extensive Plan. Such Persons, as incline to become Members, are requested to signify the same to the Secretary, by Letter, who will enrol their Names, as such, upon their paying any Sum not less than a Guinea, into the Hands of the Treasurer. The Secretary will carefully lay before the Society every Communication he may receive. Information, from Gentlemen in the neighbouring Provinces, upon such matters as they think conducive to the general Design of this Institution, will be gratefully acknowledged."[50]

While the stated objective of all of the societies was the improvement of agriculture, the areas of primary interest were quite varied and the activities of the societies differed considerably. For example, the Halifax society, in 1790, made available premiums as follows:

- " I. A Silver Medal, value one Guinea, to the person who, in the Province of Nova Scotia, shall raise the largest quantity of merchantable wheat in either the years 1790 or 1791. The Claimants of this medal must produce to the society, certificates of the respective quantities of wheat on which their claims shall be founded; and those certificates must be signed by three or more of the Justices of the Inferior Court, at one of the quarter sessions held in the counties respectively where the claimants reside.
- II. A Silver Medal, value one Guinea, to any person who shall, between May 1, 1790 and May 1, 1792, bring to the market of Halifax for sale, the fattest ox, or any other of the neat kind, whose four quarters shall weigh most, and which has been raised and fattened in the Province of Nova Scotia. The candidates for this medal must produce to the society certificates of the weight and quantity of their respective cattle, and signed by the clerk of the

market in Halifax.
- III. A Silver Medal, value one Guinea, to any person who shall between this time and May 1, 1792 produce to the society the best account in writing of the use of Plaister of Paris, as a manure for grass or grain. The society expect that the above will contain — 1. Directions for the best and cheapest methods of preparing the Plaister of Paris, by burning or grinding, 2. Information of the kind of soil to which it is best adapted, either for grass or grain. 3. Information about the quantity of Plaister of Paris per acre, best suited to grass or grain, and in different soils. 4. The properest season for laying it on the ground, and the subsequent treatment of the soil, to make it most productive in grass or grain. The claimants of this medal are to send their papers sealed under cover, and directed to the secretary of the society, not signed with their names, but dated from the village, or township and county in this province where they respectively reside."[51]

The oldest agricultural society in the province, in terms of continuity, is one formed in 1789, at Horton in King's county, as it is still functioning. It had for its purpose: "the better improvement of Husbandry, encouragement of Manufactories, cultivation of social Virtues, acquirement of useful knowledge, and to promote the good Order and well being of the Community to which we belong."[52]

This society imported seed and livestock, laid down regulations for marketing, provided an agent in Halifax to look after the sale of members' produce, carried out experiments, started a circulating library, instituted Sunday schools and in many ways tried to live up to the intention of the founders. It is evident that the society was concerned with the agricultural community in the broadest sense.

Another example of early activity was a long article from the Hants agricultural society dealing with the proper procedures for building "About-eau", draining and cultivating marshlands and cultivation and rotation of crops on uplands. Some of the details are of interest. With reference to the marshland it was stated: "it may be ploughed in the autumn, very thin, in narrow ridges; not exceeding ten small furrows in high or creek marsh, nor more than eight in low ground, and it should be immediately cross drained, so effectually as to prevent any water from standing in the dead furrows, or on any part of the ploughed field. Marsh so prepared, may be sown the following spring as soon as the surface is thawed sufficient to afford a covering for the seed, and while it remains so hard frozen below, as to bear the cattle at harrowing. From such early sowing the best crops may be expected, and an early harvesting of well ripened, full, weighty grain."

"In cultivation of upland, it will be found to be a successful process, to plant potatoes on green sward as a preparation for wheat. Less manure will be required than on other land; the potatoes will be better; and the crop abundant; and a greater certainty of good clean wheat after

The Trail Blazers of Canadian Agriculture

it."

"Upland manured with marsh, or river mud, taken from the beds or banks of rivers, on the sides of which marshes are formed, may be expected to produce good crops of wheat for two years, peas the third year, and wheat again the fourth year (the second and fourth crops of wheat the best) and if then laid down with red clover and timothy, the lasting effect of such manure will be found in great burdens of grass for many years longer than from barn door manure..."[53]

Another example of early activity is that the West River agricultural society held a plowing match in the fall of 1818.[54] Of interest here is the fact that the winners were all listed as servants of landowners.

From the foregoing it can be seen that a number of agricultural societies had been formed and were functional prior to the appearance of Agricola's letters, but the letters did stimulate interest and there was a great proliferation of local societies at that time. However, in many cases the interest was short-lived and many of the societies succumbed after a few years. In addition to the local societies, a Central Board or Provincial Agricultural Society was formed and it was through this Board that the government provided subsidies to the societies. The societies generally attempted to encourage improvement through the importation of improved seeds and livestock as well as through meetings and the provision of reading material that could add to the knowledge of farmers. Some of the societies and the Central Board imported new equipment and held demonstrations of such new equipment. This equipment also was used as a stimulus to local industry as can be seen from an advertisement that appeared in the Acadian Recorder under the heading "Provincial Agricultural Society": "As certain implements of Husbandry, consisting of Ploughs, Drill Barrows, Weeding machines, etc., models of which are in possession of the Secretary, and as it has been determined, that these should be fabricated in the Province for the encouragement of our domestic industry, this is to give notice, that sealed tenders, addressed to the Committee, and specifying the price of each article, will be received until the 20th of next month, and the lowest offer then preferred."[55]

Another activity thought to stimulate the industry was the holding of fairs. The first of these ever held in Canada was staged at Fort Edward Hill, in the township of Windsor, on May 21, 1765, long before the advent of the agricultural societies. It is of considerable interest to note the announcement of the fair and the prize list and compare them with the modern situation.

"Whereas it is thought the Establishing of the Fair at Windsor will be of great utility to the Province of Nova Scotia, a number of Gentlemen of Halifax being desirous of promoting every measure that may conduce to the public good have entered into a subscription for Premiums and Rewards and will cause the following to be given on Tuesday the 21st of May 1765, the first day of the Fair

To the person who shall bring to the Fair for sale the greatest number of Neat Cattle	3 yds of English Superfine Broad Cloth & a Silver Medal
Ditto the greatest number of horses	A saddle bridle & a Medal
Ditto the greatest number of Sheep	A pr Shears a pr Cards and a Medal
Ditto The largest pr of working Oxen	a plow Share & a Medal
Ditto The next largest	a plow Share
Ditto The finest and largest Cow	a Butter Churn and Medal
Ditto the largest yearling Bull	a Medal
Ditto do do Heifer	a Medal
Ditto do do a pair of Steers	a Medal
Ditto The best pair of Harnest Horses drawing a breast	a plow share and a Medal
Ditto The next pair of ditto do	a plow share
Ditto The best single Horse	a Medal
Ditto The next best do	a whip and a pair of Spurs
Ditto The largest and fattest wether sheep	a pr of wool cards & a Medal
Ditto The best Butter not less then 12 lb.	6 yds Ribbon & a Medal
Ditto The best Cheese not less than 12 lb.	6 yds Ribbon & a Medal

The day after the Fair will be given for the following Diversion

The best running Horse	a pair of Buckskin Breeches & a Medal
The next best do	a pair Buckskin Breeches
The next best do	a Whip and pr Spurs
To the best wrestler under 25 years old	a Laced Hat and a Medal
To the next best do	a pr Shoes and Buckles
To the next best do	a pr Buckskin gloves
To the person that shoots at a mark with a Single Ball the best in Two Shots at 80 yards distance	a Neat ( ) and a Medal

Proper persons will be appointed to adjudge the different Rewards and to regulate the diversions

The Medals will be of Silver the Size of a Dollar with an Inscription thereon suitable to the occasion

Three or Four of the best Stalions in the Province will be provided Gratis

Many of the subscribers and others will be at the Fair to purchase a great number of Cattle.

Halifax the 29th March 1765."[56]

Even at that early date a fair apparently could not be successful without having sports or entertainment events associated with it.

Later on, fairs were held under the auspices of the agricultural societies with money prizes supplanting the prizes in kind. It is of interest that as late as 1819 breeds of livestock were not mentioned in prize lists with the exception of one case where Southdown sheep were mentioned.

The newspapers of the day were instrumental in providing information of interest to farmers. Much of this took the form of articles borrowed from British or American sources and often of doubtful value in the particular environment of Nova Scotia. Nevertheless, much of it could be of value if applied with judgement. The *Acadian Recorder* was one of the most active in this field and regularly had an agricultural column. It was in this paper that Agricola found his outlet and apparently had encouragement from the publisher of the paper.

Government intervention in the agricultural industry, which is a fact of life today, had very early beginnings in Nova Scotia. The minutes of the Council during the early years of English possession indicate that agriculture *per se* did not receive much attention but they do provide glimpses of some of the problems that were of concern. An order-in-council, signed by L. Armstrong in 1733, names overseers and then gives instructions as follows: "Give notice to Every Person or Persons Residing in these Said places whether English or French that they are not to put into any of these said Flocks or Herds any Sheep Lambs or other Cattle without first appraising one or both of Said overseers with their Marks nor upon any pretence whatsoever to take any from thence without first applying to one or both of them to get the Flocks together and to be present at such place as they the Overseers may judge proper for that Purpose upon penalty of paying double the Value of the Beast taken whereof one half to be Given to the Informers and the other to the poor and also upon the penalty of paying for all Such Beast as may happen to be lost besides their loss of Commonage. And in consideration of their trouble in Attending all & Every person or persons who may have any Sheep Lambs or other Cattle in the said Herds or Flocks are (as it hath been proposed by the Deputys of the French Inhabitants & approved of in Council) to pay to them the said Overseers Six pence for Each Bullock or Cow and four pence for each Sheep..."[57]

We can see from this order-in-council that branding or marking animals was an established practice but now was being given official recognition and recording.

Lack of proper control of animals apparently was a persistent problem and in 1735 the Council minutes record: "... and also having received a Complaint from Mons. Degodalie & other Inhabitants in relation to trespasses daily Committed by Cattle through the Insufficiency of their dykes and fences he asked their Opinion whether he should not also

Issue out an order for keeping said fences etc. in good repair, under pain of a Certain penalty to be paid by the person or persons offending to those who may be Injured through such willful & Careless neglect, which being all agreed to, Orders were therefore Issued."[58]

Legislation Enacted

After the settlement and development of Halifax and the later increase in settlement in other locations, acts and regulations affecting agriculture increased. Reference has already been made to the bounties that were instituted in 1752 to encourage the clearing of land and increasing the production of grains and grasses. This was the beginning of what was to become a steadily increasing intervention by the government. The selected list of acts and regulations that follows shows that they have been both encouraging and restricting as appeared to be necessary at any given time.[59]

1758
- An Act for granting Bounties and Premiums on the fencing and improving of lands, raising grain, roots, hay, hemp, flax, and catching and curing fish.
- An Act for preventing trespasses. This act defined proper, legal fences and provided for redress of damages caused by animals trespassing on lands properly fenced.

1759
- An Act for extending the bounty on Stone Walls built and Hay raised within the peninsula of Halifax.

1768
- An Act to prevent the malicious killing or maiming of Cattle. This applied to all classes of stock.

1769
- An Act to prevent for a limited time, the exportation of Wheat, Rye, Barley, Flour, Meal, And Pease from this Province. This act was repeated on several occasions in later years, usually because of unfavourable harvests and shortages of local supplies.

1779
- An Act to prevent the spreading of distemper among horses and cattle.
- An Act for establishing a public market for the sale of livestock within the town of Halifax.

1783
- An Act for establishing the standard weight of grain, and for appointing proper officers for measuring grain, salt, and coals and ascertaining the standard size of bricks. This act was permitted to expire and was later replaced (see 1792).

1791
- An Act to prevent the growth and increase of thistles on lands in this Province.

1792
- An Act to revive, and amend, An Act establishing the standard

weight of grain...

Wheat	shall	weigh per bushel	58 pounds	avoir-dupoise
Rye	do	do	do	56 pounds	do do
Indian corn	do	do	do	58 pounds	do do
Barley	do	do	do	48 pounds	do do
Oats	do	do	do	34 pounds	do do
Pease	do	do	do	60 pounds	do do

It is interesting to note that potatoes were not included in the products listed.

The above weights were standard until 1838 when there was a revision as follows:

Wheat	60 lbs.
Barley (foreign or imported)	52 "
Barley (home grown)	48 "
Rye	56 "
Indian corn	58 "
Oats	34 "

1794
- An Act for the preservation of sheep.

1802
- An Act for the appointment of inspectors of butter in the county of Cumberland.

1805
- An Act for granting two thousand pounds for the encouragement of agriculture in the province.

1806
- An Act to encourage the raising of bread corn on new land. This act was extended in 1811.

1819
- An Act for the encouragement of agriculture, and rural economy, in this Province.

1827
- An Act relating to common fields. This in effect was a brand act.

1832
- An Act to encourage the importation of improved breeds of cattle into this province. This provided for £300 per annum for three years with the inhabitants of counties and districts to raise equivalent amounts and the monies to be used to import one bull and one cow of improved British or Irish breeds. This act was extended in 1836, 1839 and 1840.

1837
- An Act for the encouragement of the Nova Scotia Horticultural Society.

1845
- An Act to incorporate agricultural societies.

1847
- An Act to incorporate the Nova Scotia Horticultural society.

In the various acts for encouragement of agriculture two crops were given special attention. One of these was bread grains, in an attempt to attain self-sufficiency, and the other was hemp, a crop of particular interest to the naval authorities of the mother country. Hemp was mentioned in one of the earliest acts providing for bounties and premiums. It crops up again on several occasions. For example, the lieutenant-governor's opening speech to the Assembly in 1801 stressed the need for an increased production of hemp. The following year emphasis was given to this by substantial bounties, namely: "sixteen shillings for every hundredweight of found, well-cleaned merchantable hemp to the market at Halifax within the year 1803."[60] There was also offered a premium of 50 pounds to the person who delivered to market the largest quantity of clean, merchantable hemp, not less than five tons, in 1803. Thirty pounds was offered for the second largest amount, not less than three tons, and 20 pounds for the third largest amount, not less than two tons. Despite all of the inducements offered from time to time, hemp production never did become popular and did not increase as desired. It may be noted that similar efforts were made in other provinces with no greater success.

Subsidies for bread grains were often tied to newly cleared land. For example, the act of 1806 provided for the payment of ten pence per bushel of wheat and seven pence half penny for rye produced on new land. This was increased to one shilling and nine pence, respectively, in 1811.[61] Self-sufficiency in bread grains proved to be an elusive target and never was achieved consistently.

In view of the importance of fruit production in Nova Scotia at present, it is interesting to note that a Horticultural Society was formed quite early in the 18th century and was incorporated by the provincial government in 1847. However, fruit production does not appear to have achieved prominence until after that time. Eaton provides us with a brief summary of some early developments in one area of the province:

"The first fruit gardens of King's were planted by the Acadians, and a few individual apple trees at Gasperau, Grand Pre, and Canard still stand, which are supposed to have been planted by these fruit-raising pioneers. Though the first plum trees have long since disappeared, some varieties of this fruit are still grown which are traceable to these French gardens. These patches of fruit trees planted by the French encouraged the New England settlers who came in 1760 to the farms of the Acadians, and they soon began to enlarge the orchards and introduce new varieties of fruit. We have the names of several men of the early part of the century who took special interest in fruit, and we have also the names of a number of varieties of apples, some of them still standard, which these men introduced. Col. John Burbridge has the credit of having started the Nonpareil and English Golden Russet; Bishop Charles Inglis introduced the Bishop Pippin or Yellow Bellefleur; Ahira Calkin the

The Trail Blazers of Canadian Agriculture

Calkin Pippin and Calkin's Early; David Bent brought from Massachussetts the Greening Spitzenberg, Permain, and Vandevere; but one man who exerted, perhaps, the greatest influence on the early history of the industry was the Hon. Charles Prescott, who removed from Halifax to Stair's Point in 1812. Here, in his beautifully kept garden, Mr. Prescott planted the Ribston, Blenheim, King of Pippins, Gravenstein, Alexandra, and Golden Pippin, which he imported from England, the Baldwin, Rhode Island Greening, Esopres Spitz, Sweet Bough, Early Harvest, and Spy, which he obtained from the United States, and the Fameuse, Pomme Gris, and Canada Reinette, which he got from Montreal. To Mr. Prescott's credit, too, is the introduction of many standard varieties of plums, pears, and cherries since grown in the province."[62]

With reference to acts of the Assembly, certain acts were omitted that provided for the importation of grain, flour, and meal, and for the relief of poor settlers following crop failures of which occurred in 1817 and 1837. There was no mention made of the cause of the crop failures, but they may well have been caused by crop diseases or insects, or both. The earliest records make no specific reference to insect, disease, or weed problems but it is evident that by the end of the 18th century weeds were of considerable concern as the Assembly saw fit to take action for the control of thistles in the act of 1791. The seriousness of this problem is indicated by the action taken by the General Sessions of Yarmouth county in 1822 when it ordered: "That for every thistle that shall be allowed to ripen the seed the owner of the land whereon such thistle shall grow shall pay a fine of sixpence for one thistle, for two thistles a shilling, and six pence for each additional thistle until the sum amounts to twenty shillings and no more. This is to include the middle of the highway fronting said land."[63] Other weeds were a problem as Agricola and other commentators make specific reference to them.

In the 19th century insects and plant diseases became problems of major concern. It has not been possible to determine precisely when they first appeared but mention was made of smut as early as 1819; however, it seems safe to assume that it had affected crops much earlier than that. In 1834 mention was made of the "ravages of the fly" causing a reduction of 50 per cent in the wheat crop.[64] The year before a report stated that: "Blight and mildew are almost unknown to the eastward"[65] indicating that these diseases of wheat had been experienced in some localities.

While the early farmers may not have had serious problems with weeds, insects, and crop diseases they did have other problems. A major one was the occasional destruction of the dykes that protected the marshlands from the sea. Abnormally high tides and storms could breach the dykes and let the sea water in to destroy the crops and make the land unsuitable for crop production for several years.

Forest fires also constituted a problem and the *Royal Gazette* of July 10, 1792 gives an example of this: "By a gentleman lately arrived from

Shelburne, we learn that by means of the late fires which raged to a great degree in the Woods and County adjacent, there have been fifty Farm Houses consumed, together with the Fences and Crops of Grain etc.."[66]

At a later date, 1819, in a letter to Agricola, we find reference to another problem: "The mice did us serious injury; and I am mistrustful of them this season. The snow has so long covered the ground, that I fear they will come out like locusts on us in the spring and summer, and destroy the fruits of the earth. I wish you, or some of your correspondents would turn your attention to these vermin, and contrive some effectual means of guarding the country from their ravages, which in my opinion resemble so much the plagues of Egypt, that I sometimes think they are sent to us for our sins."[67]

Earlier in this chapter some statistical data were presented but it seems to be desirable, in bringing this chapter to a close, to present a tabulation that will indicate development during the period under review and to show the recent status of agriculture in the province. Data for the period of the French regime and well into the English regime are fragmentary. It was not until well into the 19th century that fairly regular and complete statistics became available. Data for selected years are presented in Table IV.

These data show that there was considerable growth and development of the agricultural industry up to 1861, as measured by acreage cultivated, numbers of livestock, and production of the various crops. However, in the course of the following century the reverse situation took place and there has been a marked reduction in practically all of the items listed.

Table IV
Periodic population and agricultural statistics for Acadia and Nova Scotia during early development, with 1966 data for comparison[a]

		1671	1686	1701	1827	1851	1861	1966
Population	No.	441	885	1134	123,630	276,854	330,857	756,039
Cultivated[b]	"	429	896	1136	292,009	799,310	971,816	485,859
Horses	"	--	--	--	12,951	28,789	41,927	5,739
Cattle	"	866	986	1807	110,818	243,713	262,297	147,636
Sheep	"	407	759	1796	173,731	282,180	332,653	38,827
Swine	"	--	608	1173	71,482	51,533	53,217	57,499
Wheat	Bu.	--	--	--	152,861	297,157	312,061	32,000
Barley	"	--	--	--	--	196,097	269,578	91,000
Rye	"	--	--	--	--	61,438	59,706	--
Oats	"	--	--	--	--	1,384,437	1,978,137	1,293,000
Buckwheat	"	--	--	--	--	170,301	195,340	--
Peas & Beans	"	--	--	--	--	21,638	21,333	--
Corn	"	--	--	--	--	37,475	15,529	--
Potatoes	"	--	--	--	3,278,280	1,986,789	3,824,814	858,000
Roots	"	--	--	--	--	499,452	642,045	23,000
Hay	Tons	--	--	--	163,212	287,837	334,287	428,000
Butter	Lbs	--	--	--	--	3,613,990	4,532,711	3,162,000

a. Data up to and including 1861 are from *Census of Canada* 1665-1871. Ottawa, Taylor, I.B. 1876, Vol. 4. Data under the heading "1966" are from the 1966 census for the first six items. The remainder are 1965 data from the 1967 *Canada Year Book*.

b. This measurement is in arpents for the first three years and then in acres.

Chapter III
New Brunswick

Information on the beginnings of agriculture in what is now New Brunswick is more fragmentary than that for Nova Scotia. Both areas were originally included in what was called Acadia and the early major developments in Acadia centred in what is now Nova Scotia.

The first attempt at agriculture on the mainland of the Maritime provinces was in New Brunswick and not in Nova Scotia, though Port Royal is usually accorded this credit. The first settlement, in 1604, by de Monts, was on an island in the St. Croix river. As this is now American territory it usually is excluded from consideration as a Canadian development. However, Champlain reported: "Afterwards some gardens were made, both on the mainland and on the island itself, wherein many kinds of grain were sown which came up very well, except on the island where the soil was nothing but sand in which everything was scorched when the sun shone, although great pains were taken to water the plants."[1] Champlain's maps show gardens on both banks of the river so there was agricultural activity on what is now Canadian territory.

There seems to be no detailed record of the crops actually sown though vegetables were included as well as grain. Rye was one of the grains for Lescarbot wrote, following a visit to the site at a later date: "As for the nature of the ground, it is most excellent and most abundantly fruitful. For the said Monsieur de Monts, having caused there some pieces of ground to be tilled and the same sowed to rye (for I have seen there no wheat), he was not able to tarry for the maturity thereof to reap it; and notwithstanding, the grain, fallen, hath grown and increased so wonderfully that two years after we reaped and did gather of it as fair, big, and weighty as any in France, which the soil had brought forth without tillage; and yet at this present it doth continue still to multiply every year."[2]

The settlement on the island was short-lived as it was moved to Port Royal in 1605. We then hear nothing further of occupation of New Brunswick territory until 1610 when a short-lived trading establishment was located on what is now Caton's Island in the St. John River but no reference to agriculture was made. The same is true of other fishing, trading, and missionary posts at Miscou, Nepisiquit, and on the St. John River, though agricultural activity cannot be ruled out completely. The first reference to such activity is that reported by Denys who had an establishment at Miscou in 1645. Of this he wrote: "The land is sandy, but nevertheless good. All kinds of herbs thrive well, and when

The Trail Blazers of Canadian Agriculture

I had an establishment there, I planted many nuts of Peaches, Nectarines, and Clingstones, and of all kinds of nut fruits, which came up marvellously. I also had the Vine planted there, which succeeded admirably."[3]

A little later, possibly 1652, Denys was at Nepisiquit of which he wrote: "It is here that I have been obliged to retire after the burning of my Fort of Saint Pierre in the Island of Cape Breton. My house there is flanked by four little bastions with a palisade, of which the stakes are eighteen feet in height, with six pieces of cannon in batteries. The lands are not of the best; there are rocks in some places. I have there a large garden in which the land is good for vegetables, which come on in a marvellous way. I have also sown the seeds of Pears and Apples, which have come up and are well established, although this is the coldest place that I have, and the one where there is most snow. The Peas and Wheat come on passably well; Raspberries and Strawberries are abundant everywhere."[4]

So, while there were some minor agricultural activities during the 17th century it was not until near the end of the century that continuous agriculture took hold. It is difficult to pinpoint geographically the exact area where this occurred. In the Nova Scotia account it was noted that, about 1672, Jacques Bourgeois established a settlement at Beaubassin. This was so successful that it attracted the attention of de la Valliere from Quebec who was granted a seigneurie of about 1,000 square miles, which included Beaubassin, much of Chignecto Bay, Cape Tormentine, and Bai Verte. Thus, it covered part of both Nova Scotia and New Brunswick. "La Valliere established himself on an 'island' of higher ground in the marshes between the Fort Lawrence and Fort Beausejour ridges (known as Tonge's Island today)".[5] This island is in present-day New Brunswick.

However, this general area attracted attention because of the extensive marshes and again it was an emigrant from the Port Royal area who spearheaded a new settlement. In the spring of 1698 Pierre Thibaudeau, with four sons and one of their friends, sailed up the Petitcodiac River to explore the territory. He chose a site on the Shepody River and then returned to Port Royal with two of his sons for supplies while the rest of the party stayed on to start building shelters. During the summer he returned to Shepody with grain, tools, two oxen, a horse, and seeds and with additional settlers. Work went on apace, but when fall came all of the men and animals returned to Port Royal for the winter. In 1699 they returned to Shepody and from then on there was continuous occupation.

In the spring of 1700 Thibaudeau, after personally having wintered at Port Royal, went back to Shepody with "The equipment for two mills, with a horse, cows, a bull, pigs, fowl and other necessities for the infant colony."[6]

The colony grew and others were established in the same general area. Thus: "In 1702 there appear to have been seven households at Shepody (roughly thirty-three people) and five households (thirteen people) on the Petitcodiac, most likely near present Hillsborough. By 1707 fourteen families and seven *engages* (fifty-five people in all) were estimated, with twelve horses, seventy cattle, and fifty sheep."[7]

Meanwhile there had been some developments on the St. John River as shown by a letter from Villier to Frontenac in 1696. "I informed you last year, Monsieur, by the memorandum I had the honor to send you, that the inhabitants living on the river had begun to cultivate their lands. I have since learned that they have raised some grain. Mon. de Chaffours, who had sown very considerably last year, has not reaped anything, the worms having eaten the seed in the ground; Mon. de Freneuse, his brother, has harvested about 15 hogsheads of wheat; Mon. de Clignacourt, very little; Mon. Bellefontaine about 5 hogsheads; the Sieur Martel very little, for he has only begun to cultivate his land during the last two years. The other inhabitants have raised only a little Indian corn. The Sieurs d'Amours, except Sieur Clignacourt, have sown this year a good deal of wheat and the Sieur Bellefontaine also; the Sieur Martel some rye and wheat and much peas. The other inhabitants have planted some Indian corn, which would have turned well only that they have sown too late their ground having been inundated."

"As the brothers Louis and Mathieu d'Amours may be considered to have been the first farmers on the St. John River we shall venture to give, from the census made in 1695, the modest figures that show the quantity of land they had under tillage, the number of their domestic animals and their crops for the past season."

"The Sieur de Chaffors had at Jemseg 65 acres under cultivation, a house, barn and stable & 22 horned cattle, 50 hogs, and 150 fowls. It is stated that his stock of animals was smaller than usual by reason of the supplies he had furnished to the French privateers. His crop of the last season (1694) included 80 bushels of wheat, 100 of peas, 30 of Indian corn, and 18 of oats."

"The Sieur de Freneuse had 30 acres of ploughed land and 40 acres of marsh, a house, barn and stable, 10 horned cattle, 40 hogs, 86 fowls. He had raised in the last season, 50 bushels of wheat, 40 of peas, 120 of Indian corn, and 12 of oats."[8]

There is no detailed description of the methods of cultivation in use but it seems logical to assume that the plow was used as the acreages listed could hardly have been cultivated with only the most primitive of tools, i.e. the spade and the hoe. Unfortunately, no good description of the cultural practices in effect during the 17th century is available.

The settlements on the St. John had a precarious existence because of the raids by New Englanders and, apparently, were not continuous.

Those on the Petitcodiac continued to grow though they too endured some hardships because of the conflict between the French and the English. "By 1734; when an ecclesiastical census was taken, there were 65 families distributed on the Shepody, Petitcodiac, and Memramcook Rivers. In the next sixteen years, more rapid progress was made, and the 1750 census attributed 170 or 165 families to the region."[9]

Then came a period of disruption with the expulsion of the Acadians and a very slow replacement of them by people from other areas. The elimination of the Acadians was not complete for the census data of 1766 show: "a population of 159 in Hopewell, two English, 12 Irish, 62 Americans, 59 Germans, 24 Acadians, these last, we may conjecture, inhabitants of Chipody who had remained in the vicinity since the destruction of their settlement."[10]

Developments continued in this area and we get a picture of a relatively prosperous community: "At Hillsborough, for instance, according to a return made in 1775, every settler had at least one cow; several of the herds numbered fifteen or more; and one herd boasted forty-eight head of cattle. All but two of the thirty-four households possessed oxen, and eleven of them reported five yoke or more. All but two establishments reported crops of potatoes and turnips, in amounts ranging from 22 bushels to 150; most of them raised wheat and oats; a smaller number reported crops of peas, barley, and rye; seventeen had raised flax."[11]

This seems to be the first mention of potatoes grown in the province. As far as can be determined the Acadians did not grow this crop prior to the English occupation. It has not been possible to pinpoint the exact time of the first entry of this crop into any part of Acadia but the first record of its introduction into Nova Scotia was at the time of the Halifax settlement in 1749.

There seems to have been continued development in this same general area for a report from Cumberland in 1785 stated: "Neat Cattle may be spared from the County of Cumberland this year to the amount of six hundred head, and eight hundred for the year 1786. From a hundred and sixty to two hundred Horses can be also spared yearly, with oats to the amount of two or three thousand bushels for their use.... As to prices they are fluctuating and unsteady. Good beef by a single cow or ox is now sold for 3d. per pound; in large quantities it will of course be less. Tollerable good horses from 12£ to 15£, others from 8£ to 12£."[12]

Following the expulsion of the Acadians new developments took place on the St. John River. McNutt states that New England soldiers who saw this area during service in the Seven Years War were impressed with its potential. "Responding to demand, the Nova Scotia authorities laid out a township of 100,000 acres in 1761, and two years later several hundred farmers from Essex County in Massachusetts occupied this garden area of the middle river. The lands were rich but the settlers poor. Carts, cattle, and grindstones were owned in common because individuals could not afford to buy them."[13] This was the settlement of Maugerville between St. John and Fredericton.

That this settlement prospered during the next 15 years is indicated by the following: "We learn from James White's memoranda that at the time of the arrival of the troops at Fort Howe he made a trip up the river to Maugerville, where in the course of five days he bought nine yoke of oxen from as many settlers on terms similar to those contained in the following statement:

Maugerville, November 16, 1777

I promise to deliver to James White, on his order, two oxen, coming five years old, when the ice is strong sufficient to bear them to drive to the mouth of this River, said White paying me on delivery fifty-five dollars.

Witness my hand, Asa Perley."[14]

Slightly later a more detailed picture of one operation at Maugerville is provided in a letter to the editor of the *Royal Gazette*: "Having heard much of the spirited husbandry of Capt. George Hardins of Maugerville, I took the liberty the other day, as I was passing through that town, to call upon him and satisfy myself. I found him very obliging and willing to gratify my curiosity — his farm is like others in that town, forty rods wide and extends from the river northerly. He has under actual improvement and cultivation about 50 acres, the following is a list of his horses, cattle, etc., viz.

1 Horse	2 last Spring Calves
1 Colt a year and a half old	49 Sheep
6 Oxen	14 Fat Hogs
6 Cows	25 Store Hogs
18 Young Cattle 2 years old	And a great variety of Poultry last spring.

I also took an exact account of the present years crop or produce of his farm which is as follows viz.

40 Tons of excellent Hay	40 Bushels of Oats
200 Bushels of Wheat	10 do Beets
300 ditto Indian Corn	5 do Carrots
20 do of Peas	50 do of Turnips
9 do of Beans	500 do of Potatoes
100 lb. of Flax	500 heads of Cabbage."[15]

The writer of the letter then went on to chastise the more indolent land owners who were unsuccessful and blamed the land for their lack of success.

With the end of the American Revolutionary War there was a major influx of Loyalists and new settlements were developed in various parts of the province. The settlers encountered the usual hardships of opening up new territory and many of them moved to other regions. Nevertheless, their arrival heralded major development in agricultural activity and production. A good description of the situation was given by Lord

Edward Fitzgerald, in 1788:

"The country is almost all in a state of nature as well as its inhabitants. There are four sorts of these; the Indians, the French, the old English settlers, and now the Refugees from other parts of America; the last seem the most civilized."

"The old settlers are almost as wild as Indians and lead a very comfortable life; they are all farmers, and live entirely within themselves. They supply all their own wants by their contrivance, so that they seldom buy anything . They imagine themselves poor because they have no money, without considering they do not want it; everything is done by barter, and you will often find a farmer well supplied with everything, and yet not have a shilling in money. Any man that will work is sure in a few years to have a comfortable farm; the first eighteen months is the only hard time, and that in most places is avoided, particularly near the rivers, for in every one of them a man will catch enough in a day to feed him for a year. In the winter, with very little trouble, he supplies himself with meat by killing moosedeer; and in the summer with pigeons, of which the woods is full. These he must subsist on till he has cleared enough to raise a little grain, which a hard working man will do in the course of a few months. By selling his moose skins, making sugar out of the maple trees and by a few days work for other people, for which he gets good wages, he soon acquires enough to purchase a cow. This, then, sets him up, and he is sure in a few years to have a comfortable supply of every necessity of life. I came through a whole tract of country peopled by Irish, who came out not worth a shilling, and have all now farms worth (according to the value of money in this country) from £1,000 to £3,000."[16]

One may be inclined to accuse the author of being overly enthusiastic in his description of the progress made and the financial status of the settlers. Nevertheless, another account, from the same year, gives some support:

"The quantity of Cattle and Grain raised will certainly be great, both for home consumption and exportation, at present many Farmers having from sixty to one hundred head, besides raising from five hundred to one thousand Bushels of wheat."[17]

A general picture of the agriculture in the province early in the 19th century comes from two sources. The first is a report on King's county in 1803:

"The productions are such as are common to America, viz. Wheat, Rye (both summer and winter), Indian Corn, Barley, Oats, Peas, Beans, Flax, Potatoes and every kind of vegetables in perfection and abundance. Horses, horned cattle, sheep, hogs, butter and cheese, masts, spars, ton-timber, staves, trenails, boards, birch plank, etc. This County is principally agricultural and annually sends about 200 or 300 barrels of flour to market, formerly more, but the ravages of the Hessian fly have considerably reduced exportation."

"The County, like most others in the Province is not in a high state of cultivation, but is making rapid progress in many parts of it, particularly at the head of Kennebecasis River where the intervales are extensive, which are generally cleared of the timber and in most places free from stumps so as to admit the introduction of the plow with facility and are inexhaustibly productive."[18]

A little later Fisher gave a fairly detailed description: "The domestic animals in this Province are much the same as those in the United States; many of the horses and oxen used in the lumber business, being annually furnished by Americans. The breed of horses has been improved by stallions imported at different periods from England and other places. In Cumberland the inhabitants have paid considerable attention to the improvement of the breed of horned cattle; in consequence of which, and the extensive marshes in that county, their dairies are superior to any in the Province. The sheep and swine are of a good size and various breeds. As Agriculture has been much neglected in this Province on account of the great trade that is carried on in lumber, not much attention has been paid to improving the domestic animals, till of late, a Society has been formed, and cattle exhibitions instituted, which no doubt will soon make an alteration in that part of the rural economy of the Province."

"The domestic Fowls are Turkies, Geese, Ducks, Hens, and other Poultry."

"The vegetable productions are Wheat, Rye, Oats, Barley, Maize, Beans, Peas, Buckwheat and Flax, with a variety of Roots, Grasses, and Hortulan Plants. The fruits are Apples, Plums, Cherries, Currants, Gooseberries, Cranberries, Blue and Black Berries, Raspberries, Strawberries, and small Grapes, with a number of small wild fruits. Butter Nuts, a large oily nut, Beech Nuts, and Hazel Nuts are found in different parts of the country in abundance, and in many places serve for fattening hogs; particularly the Beech Nuts, which after the severe frosts in the fall nearly cover the ground."[19]

"Wheat is sown from five pecks to two bushels to an acre, and yields from twelve to twenty-four bushels per acre. Twenty bushels is a good crop, on new land, although it sometimes produces more, when the soil is very rich and the season favourable. On old land the return is from ten to fifteen bushels per acre, the mean is about twelve. Rye is grown on inferior lands. It takes about the same quantity of seed to the acre, and gives much the same return."

"Oats are much cultivated, in this county, and generally turn out a good crop. The quantity of seed is from two to three bushels, and the produce from twenty to thirty bushels per acre. Barley is not much cultivated, although it would do well as a substitute in frosty seasons."

"Buckwheat is a grain that gives a large return for the quantity sown. It is raised on lands that are too poor to produce good crops of other grains, and sown later in the season, so that the greatest summer heat may be past before the grain is formed in the ear; for should there be a few very hot days when the grain is in the milk, the crops will be destroyed. The same would be the case, if a slight frost should strike at that stage. If, however, it escapes these casualties, to which it is liable, it turns out a good crop, yielding from forty to sixty bushels to an acre. There is a species of wild Buckwheat, which is a surer crop, but of an inferior quality."

"Millet has lately been introduced into the Province. It is said to do well on most lands, but has not been much attended to."

"Indian Corn or Maize flourishes in high perfection on the intervales, which are generally composed of alluvial soil. It is usually planted in hills nearly four feet asunder. Five grains is the usual quantity for a hill. It is a plant that requires a light soil, old manure, and hot seasons; should these requisites occur, a good crop may be expected. It is usually hoed thrice, and produces from twenty-five to forty bushels per acre."

"Pease are a hardy grain, and produce from ten to fifteen bushels per acre. Beans are usually set in drills; they thrive well on light sandy lands, but are not much cultivated in the country."

"Among the ground crops or roots, the most valuable is the Potato — a root that can never be sufficiently prized, as affording one of the most productive and surest substitutes for bread of any known, and without which it would have been extremely difficult to have colonized these Provinces. This may be reckoned the surest crop, and is peculiarly well adapted to new countries, as it thrives best on new burnt land."

"The usual and simplest method of cultivating this root is by planting cuttings of it in hills, about three feet asunder. This method is peculiarly convenient on land newly cut down, as the seed is set with the hoe between the stumps and roots with which the ground is covered, and where the plough or harrow could be of no service. They are generally hoed once in the season, and turn out in the fall a large crop of clean, smooth potatoes, of a superior flavour to those grown on old lands. The produce is from 150 to 200 bushels from an acre; although they sometimes exceed that quantity. They are an excellent crop for improving new lands; for as the culture is all performed with the hoe or hack, the small roots of the stumps are destroyed in planting and digging; for wherever there is room to drop an eye, it never fails to vegetate, working under roots and around stones, so that in the autumn the farmer has frequently to cut away or dig under roots for his crop, which often exceeds his expectations."

"In some parts of the Province, where the lands have been long in cultivation, drilling is practised, and the labour chiefly performed with the plough and harrow; and of late the Irish method of setting them in beds has been introduced. There are many varieties of this root cultivated in the Province; but no attention has been paid to renewing the seed from the ball, which no doubt would improve the quality as well as the produce."

"Several kinds of Turnips are cultivated in this Province; the best of which is the ruta-baga, or Swedish turnip. This is an excellent root and cultivated with great success, particularly in new lands. They differ from the common field turnip, being of a firm texture they keep the year round, while the common turnip turns soft and unfit for use as the winter sets in. They, however, answer a good purpose for early use and for cattle, being sown late in July, after the other crops are out of the way. The Swedish turnip is sown in early June. All the sowing in this country is broadcast, the method of drilling being scarcely known."

"The principal grasses produced in the country, are white and red clover, timothy, lucerne, browntop etc. Good uplands produce one and a half tons per acre, and the intervales from two to three tons. There are several species of wild grass, such as blue-joint, etc. found in meadows, in the woods, and along streams, which make very good food for young stock."[20]

In vegetables a considerable range of varieties was available as can be seen from the following 1817 advertisement: "Garden Seeds Among which are the following. Onion, White Portugal, silver skinned, Strasburgh, and Blood red. Cabbage, early York large Sugar Loaf, Drum Head, large Scots, and curled green Savoy. Raddish, white turnips, red do, scarlet salmon, and black Spanish. Carrot, early horn; Lettuce, tennis ball; Majoram, sweet; Parsley, curled; Parsnip, smelling; Savory, winter and summer; Thyme; Turnips; red Beet; Beans, broad Windsor.

N.B. The above seeds are from England, and all of last years growth."[21]

Timber was Dominant

Both of the foregoing accounts make reference to the lumber industry, a factor that was partly responsible for the relatively low level of agricultural practice in effect. It provided desirable employment opportunities and tended to draw available labour away from agriculture. This is explicit in a statement by the editor of the *Royal Gazette* in 1819:

"Agriculture in this Province is more in its wane than in the sister Province of Nova Scotia. This is occasioned principally by the extensive Forests of Pine, and the numerous Rivers and Streams which facilitate the transportation of the Timber to the places of exportation. Timber getting is all the rage. The young Farmers, and many of the old ones too, leave their farms to make their fortunes in the woods. We seem at

The Trail Blazers of Canadian Agriculture

present to be living upon the capital instead of the annual produce; and not only so, but the habits of the Farmers are undergoing change, which must prove, eventually highly injurious to the best interest of the Province."[22]

Despite this somewhat pessimistic view both of the industries have continued to flourish in the province. Yet, even today, we see a similar effect in areas where agriculture is carried on at the fringes of forests and constitutes a secondary industry to employment in the woods. In these circumstances agriculture often is neglected and is conducted at a low level of excellence.

Other factors also were responsible for the relative backwardness of agricultural practices. The *Royal Gazette* made reference to this:

"In common with Nova Scotia, Agriculture is also languishing for want of that encouragement and those stimulants which are afforded and given to it in other Countries. Fortunately for that Province (and for this too), they are roused from their lethargy, and are making great efforts to put the Farming Interest upon a respectable footing. We have with great pleasure, copied from the *Halifax Journal* the account of the proceedings which lately took place there, in the formation of a Central Agricultural Board. May we hope that this will contribute in some small degree to excite a spirit of emulation in this Province."[23]

The hope of the editor for action on agricultural societies in New Brunswick was soon fulfilled. In January, 1820, the Charlotte County Agricultural and Emigrant Society was formed at St. Andrews. A month later, at Fredericton, a meeting resolved that; "A Central Agricultural Society be formed at Fredericton, for the Province of New Brunswick."[24] At a subsequent meeting, March 4, final action was taken to form the New Brunswick Central Society for promoting the Rural Economy of the Province. Actually this was not the first such society in the province as one had been formed in Saint John in 1790, although this society did not survive.[25]

The new Central Society went into action at once. At its first annual meeting, on March 18, 1820, it passed several resolutions:

"Resolved, That the sum of two hundred pounds be laid out in the purchase of Seed Grain. Resolved, That one Ram, one Boar, and one Sow, for each County be imported into this Province during the ensuing season, provided the funds of the Society admit of the expense. Resolved, That a Drill Plough be imported into this Province, during the ensuing season, provided the funds of the society admit of the expense. Resolved, That the Central Committee, be and are hereby authorized to carry the above Resolutions into effect."[26]

Further action followed as can be seen from the following notices in the press: "April 25, 1820. An Expanding Harrow, suited to the Drill Husbandry possessing the combined propertys of a Horse-Hoe, a Harrow, a Garden Rake and Plow, may be seen on the Square for the next succeeding eight days."[27]

May 16, 1820: "Seed Oats from Scotland, for the Agricultural Society, will be at Fredericton by the earliest opportunity. As the Secretary has advanced payment on account of the Agricultural Society, those who wish may obtain not exceeding three Bushels each by applying at Mr. Richard's, and paying 5s cash per Bushel on delivery. Seed Wheat, Timothy and Clover, may also be had at Mr. Richard's."[28]

May 23, 1820. "From Cumberland Marshes has been procured by the Agricultural Society, some excellent Seed Wheat, which may be obtained on application to the Directors, as also some Potato Oats from Scotland."[29]

After this first splurge of activity the society seems to have become dormant, or even moribund, for in 1825 action was taken to establish a new central society entitled the New Brunswick Agricultural and Emigrant Society. This new society was "to superintend the Interests of the Province at large; to apportion to the different Counties, all monies which may be given by the Legislature in aid of the view of the Society. This Society reserving such part as it shall deem necessary for its own interests and to audit and examine the accounts of its own expenditures, as well as the expenditure of the sum apportioned to the different counties."[30]

It was stated that the county societies were to have full jurisdiction over expenditure of funds raised by them or given in grants.

A most interesting feature of the society was that all members of the legislature were to be considered as members of the society. At the second annual meeting of the society, January 7, 1826, it was reported that the legislature had agreed to appropriate £300 in aid of the society. The grant was accompanied by the suggestion that emphasis should be given to importation of seeds and improved implements and that these should be sold at cost to members or else be sold at public auction.

The Central Society swung into action immediately after its formation. In an attempt to stimulate production it offered a number prizes for 1826, having been formed too late to give effect to them in 1825. The prize list is interesting, not least because of the small acreages involved:

	£	s
"For the greatest quantity of best quality Wheat raised from one undivided Acre,	5	0
For the second best do do	2	10
For the greatest quantity of best Oats raised from one undivided Acre,	5	0
For the second best do do	2	10
For the greatest quantity best Barley from one undivided Acre,	5	0
For the second best do do	2	10
For the greatest quantity of best Indian Corn		

The Trail Blazers of Canadian Agriculture

	£	s
raised from one undivided acre,	5	0
For the second best do do	2	10
For the best quantity of early bluenose Potatoes raised from one undivided acre of upland	5	0
For the second best do do	2	10
For the greatest quantity of Swedish Turnips grown in the Province	3	0
For the second best do do	(missing)	

It is proposed to have a Cattle Show in Fredericton on Wednesday the Seventh day of September next when the following prizes for the encouragement of the breed of Horses, Horned Cattle, Sheep and Pigs will be offered from the Funds of the Society."[31]

The cattle show was held as planned but the report of the show indicated that the response had not been as great as had been hoped. Entries were limited for cattle and there were no entries for sheep.

The Saint John Agricultural and Emigrant Society also held a cattle show the same year and in addition to prizes for livestock, had a number of production prizes. They differed from those proposed by the Central Society, indicating some slight difference in emphasis. Their prizes were:

	£	s
"For the greatest number of bushels of Potatoes raised from one undivided acre of Land	5	0
For the greatest number of bushels of Barley raised from one undivided acre of Land that was in Potatoes the preceding season	3	0
For Oats in the above manner	3	0
For the greatest number of tons of Merchantable Hay, cut from one undivided acre of Land that had been in a rotation of Potatoes, or any green Crop, and of Wheat, Barley, or Oat	3	0

The following Premiums will be given on Crops raised on new Lands from the Forest:

	£	s
For the greatest number of bushels of Turnips raised on an undivided acre of land	3	0
For the greatest number of bushels of Potatoes raised on an undivided acre of Land, being the first Crop	3	0
Do do of wheat, whether it followed Potatoes or Turnips, and was sown with Hay seed	3	0
Do do of Barley	5	0
Do do of Oats	2	0
For the greatest number of tons of Merchantable Hay, being the third Year after the felling of the Forest	3	0
To the best improved Farm in all respects, of not more than six years standing from the clearing of the Forest	10	0
To the second best ditto	7	0
To the third best ditto	5	0

N.B. These three last Premiums are confined to persons who have been the actual occupants of the Land during the whole period."[32]

The Saint John Society also had a prize for "the best single mould board plough."³³

The prize lists for the livestock shows were relatively short compared to those of today. An interesting item for the Fredericton show of 1826 was the restriction on those who could show. "No Premium to be awarded to any person who shall not be a Member of the Central Society, or someone of the County Agricultural Societies, in this Province; and no owner of an Animal for which any premium may have heretofore been awarded will be entitled to any of the above premiums for the same Animal."³⁴ The prize list was:

	£	s
"For the best Provincial bred Stallion rising 4 years old	10	0
For the best pair of Provincial bred Mares or Geldings, not less than 4 nor more than 10 years old, owned by one person	10	0
For the best Bull, not less than 2, nor more than 4 years old; not being one of those intended to be imported by this Society	5	0
For the second best do do	3	0
For the best Cow not less than 3 years old do do	2	10
For the best Ram, not less than 3 years nor exceeding 5 years old, do do	5	0
For the second best do do	2	10
For the best Ewe	2	0
For the best Boar	2	0
For the best Sow	1	0
For the best fat Ox, not under 4 years old	5	0
For the second best do do	3	0
For the best pair of working Oxen not less than 4 years old, owned by one person	5	0
For the second best pair of working Oxen, do do	2	10
For the five best fat Weathers, not less than 4 years old, owned by one person	5	0
For the five second best do do	2	10
For the greatest quantity of good Butter, produced from any one Dairy in this Province, between the 1st May and the 1st November, in the present year	10	0
For the next greatest quantity of Butter, do do	5	0"³⁵

Breeds were not designated in this list and in this respect the prize list for the Saint John show a year earlier differed in that it had separate prizes for Dishley and South Down rams and ewes.³⁶ As the years progressed younger animals and various breeds were included in the prize lists. Shorthorns or Durhams were first mentioned in 1827, Devonshire sheep in 1828 and in that year four cows and one bull of the Ayrshire breed were imported.

There was quite a surge of activity on the part of the various societies during the third decade of the century. For example, in the second annual report of the Central Society it was noted that:

"The Central Society have imported implements to a large amount, and to promote the latter (i.e. livestock) they imported from England in the course of last summer, a strong and beautiful horse, and 16 Dishley or Leicestshire Sheep. Measures were taken to introduce the Sheep and implements of Husbandry into various parts of the Province, and the Horse will probably soon be offered for sale at Public Auction, in order that the Inhabitants of all parts of the province, may have a fair and equal chance of competing for so valuable an animal."[37]

The same report showed that the Westmorland Society had imported eight rams of the Dishley breed, the Northumberland Society reported that they had sent to Scotland for four cows and one bull of the Ayrshire breed, and Sunbury reported the importation of eight South Down sheep, the establishment of an annual cattle show, and ploughing matches.

One notable addition to the prize list of the Fredericton show in 1827 was "a premium of £10 to the person who shall have planted on the Farm between the Spring of 1822 and October 1827 the greatest number of Apple Trees (not less than one hundred)."[38]

Some details of activities of societies are indicated by the following two examples from the books of two different societies.

		£	s	p
1841	By Garden Seeds from Boston	3	0	0
	" Freight; duty, etc.	3	15	0
	Black oats from P.E.Island	2	0	0
	White & Grey Peas, Horse Beans	4	0	0
	Spring Vetches, 3 Sots each	4	0	0
	Turnips- Cabbage Seeds from England	4	0	0
	Agricultural Books	6	11	0
	Postage etc. of the New England Farmer		18	0
	2 Buck Sheep from Woodstock	5	0	0
	2 " " from S. Stephens (Leicester)	7	10	0
1842	Garden Seeds from Boston	3	0	0
	Balance in the Treasurer's hands	22	1	0
	(Mr. G. Harding)	54	15	0
1844 Jan 1	To Balance in plant account rendered	2	9	9
19	" cash paid Mr. Weldon for Agricultural Publications	1	0	0
Feb 1	" cash paid for 100 lbs Timothy	2	10	0
11	" cash paid for Premiums at Grain Show	11	7	0
Sep 26	" cash paid for Premiums at Cattle Show	13	17	0
Nov 10	" cash paid for 400 lb. clover, Freight of do 2/	15	2	0
Dec 31	" cash paid for 600 lb Timothy	15	0	0
	" cash paid Postage of Agricultural papers	0	5	0
	" loss on money	0	3	0
		61	14	9[39]

The societies were active in importing livestock and seeds but were not the first to take such action. Private individuals had taken the initiative with horses and in 1799 the following advertisement appeared in the *Royal Gazette*:

"To Farmers and Others. The imported high bred and full blooded Horses Eclipse and Phoenix, will cover Mares this season at 12 Dollars each, and a half dollar to the Groom. Eclipse will be kept at Fredericton, and Phoenix will remain in this city. — E.S. hopes the price will not be thought too high, as the Horses landed here stand him in £556:9:6 Sterling. St. John, 13th May, 1799."[40]

In 1816 the stallion Herod had appeared on the scene. Various factors suggest that these horses were more in the class of riding horses than heavy farm or cart horses. Privately imported seeds were advertised as early as 1816.

While the agricultural societies were the main factor in working for the improvement of agriculture, the newspapers also attempted to be of service by publishing articles and other pertinent information. This activity preceded the development of the societies and continued after their formation. Most of the published material was borrowed from English or American sources and, in some cases at least, lacked full applicability to the conditions in New Brunswick. An example of the material published is the following:

"Horses. — Horses may be kept in good order by feeding them with raw potatoes, or carrots, washed clean; though they would be more nourishing if they could be fed on them boiled. Hogs. — Boiled hay will keep hogs during winter; the addition of some potatoes, boiled with the hay, will make them thrive better. Sheep. — Feeding sheep with Indian corn, about half a gill each per day, is very beneficial and keeps the flock in good heart. Hens. — It is a fact not commonly known, that the reason why hens do not lay eggs in winter is the want of lime to form the shell. Let them have access to wheat, which contains lime itself, they will lay as well in winter as any other time. There is a town in the county of Schenectady where limestone abounds, and where the shell of eggs is much thicker than common, for this very reason."[41]

The various steps taken for the encouragement and development of agriculture could not overcome or completely alleviate the numerous problems faced by the settlers. In addition to the common hardship of having to clear new land the settlers of the 18th and early 19th centuries were subjected to the hazards of weather and to having their crops damaged or destroyed by weeds, insects, and disease. Thus, a letter to the editor, on October 17, 1786, referred to a frost that destroyed much of the crop and then went on to discuss the benefits to be derived from raising winter wheat rather than spring wheat or other grain.[42] Fisher,[43] whose work was published in 1825, believed that there had been a gradual change of climate in New Brunswick during the past century and especially since 1783. He referred particularly to 1816 when the cooling trend had reached its peak and there had been frost in every month of the year, causing serious crop losses.

Floods, Fines and Pests

Floods were regular occurrences along the rivers, especially in the spring, but one report, in October 1798, stated: "The rain last week has raised the Freshets in the River St. John, Kennebekacies etc. to a height scarcely ever remembered at this season of the year, & has done great injuries to the hay; many hundred stacks are said to be under water."[44] Another report in a later year told of the destruction of buildings and livestock by floods.

Forest fires, a hazard common to all settlers in forested areas, took their toll as well. The year 1825 was apparently a very bad year in this respect with major fires, especially in the Mirimachi area, wiping out several communities with loss of life and destruction of buildings and supplies.

It is not known precisely when insects first became an important factor but Villier, in 1696, mentioned that "the worms having eaten the seed in the ground" had caused complete failure of a crop. The Hessian fly was reported to have ravaged the wheat crop as early as 1790.[45] The wheat midge apparently made its appearance about 1841 for Johnston reported:

"In the year 1841 or 1842 the wheat in this Province began to be injured by destructive insects, having the appearance of very small yellow coloured maggots. Five or six of them were usually found within the outside covering of a single grain at the time when the crop was beginning to ripen. This single grain they entirely destroyed, without appearing to meddle with any of the other grains in the same ear. Hence in many ears a number of the grains escaped, and thus the quantity of produce was diminished without at all affecting the quality of what was left..."

"These insects first appeared in Sussex Vale, in King's county, and seem to have spread from that fertile district, as a common centre, all over the province. In 1844 they destroyed nearly all the wheat in the low grounds in that valley; on the high grounds in the vicinity their ravages were chiefly confined to the outsides of the fields, and to a comparatively small number of grains in each ear... In 1849 what little wheat was sown, when it grew up, was so injured by the rust, that their ravages could not be so well ascertained. This present year, 1849, some traces of them were found in the northern parts of the Province, but in all other places they have for the most part disappeared, and have left the wheat of this season almost entirely uninjured."[46]

The first appearance of any crop disease cannot be pinpointed. No report of smut was noted, but, considering the experience elsewhere, and the general occurrence of this disease, it most likely was present from the very earliest days of grain production in the province. Rust and mildew on grain were mentioned in 1816 and may well have been experienced before that time.

Disease was possibly of greater concern to the potato grower than to the grain producer. Johnston reported: "The Potato Disease here, as elsewhere, has confessedly paralyzed the rural industry of many districts, greatly added to the other distresses, especially of new settlers, and very much retarded the agricultural progress of the Province."[47]

The seriousness of this disease is indicated by a report from Charlotte County in 1845 :"Traces of the disease were observed in the Potatoes in some locations in 1843 and 4, although the Crops were generally good, but in 1845, the disease was general throughout the county, and may be considered a total failure as very few farmers took out of the Ground the quantity of seed planted and those in a diseased state and not expected to keep for seed. It is considered by many that the Potatoe has not recovered its original healthy condition since it was attacked with what is called dry rot which was first observed in 1832-3."[48] This report indicates that disease, possibly a different disease, had been affecting the potato crop for some time. It is interesting, and possibly significant, that Fisher, who dealt with agriculture at considerable length in his history published in 1825, makes no reference to disease in the potato crop, though he does refer to rust in the grain: "Roots come to perfection and grain gets ripe in most years; wheat being oftener hurt by rust than the frost."[49]

Weeds became a concern as early as 1797 when an act was passed by the legislature to prevent the growth of thistles. There is no evidence that this legislation was very effective. Other weeds were not mentioned by writers during the time under consideration, but no doubt they were in existence and of concern.

Legislation Enacted

The government of the day showed some concern about agriculture. After 1784, when New Brunswick became a separate province, numerous acts were passed regulating and supporting the development of agriculture. The most interesting and important of these early acts are listed below:[50]

1786
- An act to empower Justices of Sessions to regulate fields and fences, restraining swine, horses, and cattle and for preventing trespasses. An act to prevent the malicious killing or maiming of cattle. An act to lay a tax on dogs.

1792
- An act to prevent the destruction of sheep by dogs. An act to encourage the destroying of wolves.

1793-4
- An act for appointing Commissioners of Sewers.(This had reference

to dykes and drains in the marshes and not to sewers as we use the term today.)

1797
- An act to prevent the growth of thistles.

1803
- An act for regulating the exportation of butter. (This included the appointment of inspectors of butter).

1807
- An act to enable the owners of stray cattle to more easily recover the same.

1814
- An act to empower and authorize the Justices of the County of Westmorland at their General Sessions of the Peace, to regulate the grazing and depasturing of the several marshes, low lands or meadows, within the said county.

1816
- An act to provide for the punishment of horse stealing. An act to prevent the cutting or breaking down the bank of any river, seabank or dyke, and the preservation of the same.

1817
- An act to encourage the raising of bread corn on new land. An act to prohibit the exportation of corn, meal, flour and potatoes out of the province for a limited time. (This was in consequence of the crop failure of the previous year.)

1820
- An act for granting bounties on grain raised in this province.
- An act to provide for and encourage the settlement of emigrants in this province.

1834
- An act for the punishment of cruelty to animals.

In concluding this chapter a few comments by observers at the mid-point of the 19th century are pertinent. In commenting on the handling of livestock Johnston said: "Again, the winter feeding in the Colony is generally very much in the condition in which it was over a large part of Scotland some sixty years ago. To keep his stock alive was then the chief ambition of the Scottish farmer during the winter months, and he trusted to the nourishing grass of spring and summer to make up for the starving system of the colder part of the year. Such is very much the practice in many parts of New Brunswick, but it stunts the cattle in their growth, and even in a money point of view is a false economy. The working ox, when spring arrives, has not sufficient strength to do all the work which the urgency of the season requires, while the animal which is sold for beef has so small a weight of muscle and fat, compared with that of its bones, and the quality of the meat is so inferior, that it is comparatively worthless in the market."[51]

He reported the estimated yields of crops as shown in Table I.

Table I
Estimated yields of crops, in bushels per acre
(Excerpted from Johnson's Table V)[52]

	Maximum	Minimum	Average
Wheat	40	8	19 11/12
Barley	60	10	29
Oats	60	13	34
Buckwheat	70	15	33 3/4
Maize	80	20	41
Potatoes	500	100	226 1/3
Turnips	1000	200	456
Hay (tons)	4	1	1 3/4

The average weight per bushel was: "Wheat 60 3/4; Barley 50; Oats 38; Rye 52 1/2; Buckwheat 48 8/11; Maize 59 1/2; Potatoes 63; Turnips 66; Carrots 63."[53]

In dealing with the more advanced farming he noted from Brown's report to him: "At the present time, the degree of skill manifested in farming, and the extent of progress made, are more owing to casual or accidental circumstances, than to the relative advantage or natural capabilities of the land in the different Counties. Foremost in agricultural improvement stands the county of Northumberland, where thirty years ago it was confidently affirmed, that as soon as the pine timber disappeared the inhabitants would disappear also. In Newcastle, Douglastown, Chatham, and Napan, in particular, the appearance of the fields, the ploughing, the implements of husbandry, stock, buildings, fences, etc., all indicate an advancement in agricultural skill beyond what is to be found in any other part of the Province."[54]

Robb, in his overview, gives a slightly different picture when he states: "New Brunswick farmers at present generally belong to a class who are easily satisfied; almost all may raise enough for their daily wants; they have no rents or tithes to pay; they need not amass riches for their children, who can work the old farm, or take another as suits their inclinations; the log-crop is not yet exhausted, and it promises an immediate return; these influences, be they good or bad, certainly tend for the present to retard agricultural development."[55]

Looking to the future he was somewhat pessimistic: "On the whole, therefore, I candidly confess that I see little prospect of New Brunswick becoming a wheat exporting country. So far from being the case at present, there is as we have seen, a deficiency of four-fifths of what is actually required in the home market. Moreover, I am not sanguine that either wheat or Indian corn, the great food grains of North America, will ever become the general or staple crops in New Brunswick, or that they can be raised here at a price which will allow them to compete abroad with the same articles grown in New York, Ohio, or Canada West."[56]

The data in Table II suggest that Robb was correct as far as wheat and corn are concerned for the production of these crops has actually declined since his time. The livestock population has decreased along

The Trail Blazers of Canadian Agriculture

with the cultivated acreage. The only real advance in production has been in potatoes which constitute the major crop in the province at the present time.

Table II
Periodic population and agricultural statistics up to and including 1861 with 1966 data for comparison[*]

		1695	1840	1851	1861	1966
Population	No.	49	56,162	193,800	252,047	616,788
Cultivated	Acres	166	435,861	643,954	885,108	638,649
Horses	No.	--	18,282	22,044	35,347	6,104
Cattle	"	38	90,260	157,218	161,462	136,467
Sheep	"	--	141,053	168,038	214,092	28,235
Swine	"	116	71,915	47,932	73,995	34,126
Wheat	Bu.	130	--	206,635	279,775	167,000
Barley	"	--	--	74,300	94,679	117,000
Oats	"	30	--	1,411,164	2,656,883	3,461,000
Buckwheat	"	--	--	689,004	904,321	3,000
Peas & Beans	"	173	--	42,663	30,667	--
Corn	"	370	--	62,225	17,420	--
Potatoes	"	--	--	2,792,394	4,041,339	11,280,000
Hay	Tons	--	--	225,093	324,160	412,000
Butter	Lbs.	--	--	3,050,939	4,591,477	5,880,000

[*] Data up to and including 1861 are from *Census of Canada* 1665-1871. Ottawa; Taylor, I.B. 1876 Vol. V.

Data under the heading 1966 are from the 1966 census for the first six items. The remainder are 1965 data from the 1967 *Canada Year Book*.

Cultivated acres are in arpents for 1695 and in acres for the succeeding years.

Chapter IV
Prince Edward Island

Though Prince Edward Island, or Isle Saint Jean as it was first known, was discovered by Cartier during his first voyage to Canada in 1534, settlement with agricultural activity came late in relation to developments in Nova Scotia, New Brunswick, and Quebec. One of the primary reasons for this late development may well have been the lack of marsh areas, so abundant on the Bay of Fundy. We have noted the relative ease of cultivation of the dyked marshes, the high fertility of the marshland soil, and the reluctance of the habitants to clear and cultivate the upland, forested areas in Nova Scotia and New Brunswick. On Prince Edward Island there was no alternative, the only substantial area of land available was covered with forest.

In 1653 Nicolas Denys, who previously had been involved in settlements in Nova Scotia and New Brunswick, was given the island as a grant.[1] He retained it for ten years but then lost it because he failed to develop any settlement there. The island then was granted to Doublet, in 1656, but again no development ensued.[2]

It was not until after the English occupation of Acadia, following the capture of Port Royal in 1710, that settlement, with an agricultural base, first occurred, though some fisheries had been operated from the island prior to this. Apparently not willing to live under English rule, a number of Acadians emigrated from Nova Scotia to the island in 1711.[3] However, this settlement did not succeed for Governor Caulfield at Annapolis Royal reported, in 1716, "The Island of St. John is completely abandoned by the people of Annapolis who went to settle there. The people of Minas resolve to remain where they are."[4]

In 1719 a grant of the island was made to an influential Frenchman, the Comte de Saint-Pierre, who set about making preparations for colonization. In 1720, a substantial fleet, with supplies and 300 people, was sent from France to Port La Joye, on the harbour where Charlottetown is situated.[5] This was a major development and marked the beginning of continuous occupation of the site. Whether it can be claimed to be the first permanent settlement is in some doubt for: "From the census rolls from a later date we know that two Normans, Francois Douville and Charles Charpentier, had settled independently at St. Peters in 1719 and that Mathieu Thurin, a Canadian, had settled at East Point the same year."[6] However, these may well have been fishing establishments with no agricultural activity.

The Trail Blazers of Canadian Agriculture

The settlers who came to the island had one distinct advantage over the first settlers in Nova Scotia. They had available to them the knowledge and experience of the habitants of Acadia and also had relatively ready access to a supply of livestock and seed from the same source. But they were faced with the necessity of having to clear the forest before they could begin to till any substantial area of land as there was very little marshland available.

Most of the settlers who came in 1720 were from France, but a limited number came from Acadia. There was a slow but gradual increase in population and agricultural activity on the island with a total of 572 people in 1734 with 332 cattle and 119 sheep.[7] With the founding of Halifax in 1749 and increasing pressure on the Acadians from the English, there was an upsurge of immigration for a year or two. By 1753 the population had reached 2,256 with 152 horses, 2,317 cattle, 1,440 sheep, and 1,651 pigs.[8] Then came the expulsion and most of the French population was removed or moved away, though a few remained in isolated areas and formed the nucleus of an increasing French population after the English occupation.

It is difficult to get a true picture of the actual acreages of land under cultivation at any time during the French period. The small group of settlers who came from Acadia in 1710, and had disappeared again by 1716, no doubt cleared some land, but the amount would have been relatively small. The first census of the colony was taken in 1728 at which time there were 54 houses, 76 men, 51 women, 156 children, and 14 domestics for a total of 297 not counting 125 fishermen.[9] But this gives no indication of the land cultivated. The first indication of this activity came in the 1730 report of Pensens in which he stated that the habitants of Port La Joye had gathered more than 250 hogsheads of grain, those at Malpeque forty, Tracadie thirty, Havre a l'Anguile or Savage Harbour thirty, and St. Peters over fifty.[10] If these figures are converted on the basis of eight bushels to a hogshead we come up with a total of 3,200 bushels of grain, assuming a yield of 15 bushels per acre would give slightly more than 200 acres of land in crop, whereas at 20 bushels per acre the acreage would be 160.

Another indication of the extent of cultivation can be obtained from the fact that the individual settlers had seeded from less than two to as much six hogsheads of grain. Using the same conversion figures, and assuming a seeding rate of two bushels per acre, this would give individual acreages as great as 24 acres. This would not be impossible for a family to achieve in a 10-year period, especially if there were grown sons in the family; however, it would not be the norm.

In 1739 Duchambon reported that 669 3/4 bushels of wheat and 150 bushels of peas had been sown and that there was land cleared for sowing half as much again.[11] Using the above rates of seeding this would mean slightly over 400 acres of land seeded and another 100 acres cultivated. In 1742, 1,500 bushels of grain were sown,[12] which would indicate 750 acres of land in production.

More and more land was cleared but, because of crop disasters of various kinds, there often was insufficient seed to seed all of the land. Thus in 1752 only 1,817 bushels of seed were sown though enough land was cleared for sowing 2,935 bushels.[13] This would indicate about 1,500 acres under cultivation, a conservative estimate compared to the 1753 census data. That year 2,149 3/4 bushels of wheat had been sown on cleared land and 605 1/2 on burned land. Additional land was cleared for 5,555 bushels and burned land for 2,429 bushels. This would indicate 3,700 acres of cleared land and 1,500 of burned land available for seeding.[14] An additional 800 bushels, in round numbers, of other grains had been sown for a further 440 acres. The following year approximately 4,500 acres were seeded.

Because forests had to be cleared, the acreages cultivated by most of the settlers were relatively small, though a few may have been fairly extensive. Franquet[15], in 1751, reported on his visit to the island and mentioned particularly the farms of Sieur Bauthier and Sieur Bugeau, each of whom occupied a farm of 160 acres, though it is not stated that all of this was cleared. The 1752 census indicated a range of land cleared by individual settlers from nil, through enough for a garden, to a maximum (calculated from bushels seeded or to be seeded) up to about 50 acres.

Though there is no description of the cultural practices followed, and the equipment in common use, the indications are that the level of husbandry practised by the French inhabitants was not very high. Basically the pattern of husbandry in Acadia was followed with the main crops being wheat, peas, cabbages, and turnips though some barley, oats, rye, flax, and buckwheat were raised. The first mention of tobacco occurs in 1740 but the officials discouraged its production for fear that it would lead to neglect of wheat production. Whatever the reason, tobacco did not become an important crop on the island until more than 200 years later.

Prevost,[16] in 1753, commented on the inadequate agricultural practices which led to degeneration of the plants and depletion of the soil. He attributed this to the cropping program and the lack of fertilization and suggested the need for resting the land for as much as three years, a solution that would require additional clearing of land. A basic problem was that the natural fertility of the upland soils was much below that of the marshland soils cultivated in Acadia and thus, the practices that had been adequate in Acadia were not suitable for the island soils. Prevost also suggested the need for introducing new seed.

Land Clearing an Arduous Task

Reference was made to the use of the plow in 1738 and, considering the relatively large number of oxen shown in the censuses, it is clear that there was adequate power available for plowing. However, newly cleared land would be almost impossible to plow so it is likely that much, if not most, of the cultivation was done by hand with hoe and spade.

Clearing the land was an arduous task, especially with the primitive equipment available for the purpose. We get a hint of one piece of equipment used in the report on Roma's settlement: "For clearing the land he made several levers 35 feet long on a pivot of 20 feet..."[17] One advantage of forest fires was that they made subsequent clearing of the land somewhat easier.

There is no reference to the first arrival of domestic animals on the island. It may be assumed that the Acadians who went there in 1711 would have taken some livestock with them, but it would have been removed when they left. There is no record of what livestock, if any, the 1720 settlers brought with them. But we can infer. Gotteville, the leader of the settlers, in reporting on the natural resources of the island states: "The only animals lacking were dogs and horses."[18]

Reference to cattle was made in 1731, but it was not until the census in 1734 that a record was obtained of the livestock population.[19] This census listed 332 cattle and 119 sheep; in 1735 there were 433 cattle and 190 sheep recorded. Later census data, shown in Table I, show some variation.

Table I
Livestock census data for the years 1751, 1752, 1753[20]

Class	1751	1752	1753
Oxen	568	802	820
Cows	566	699	711
Bulls	192	85	184
Heifers	192	185	234
Calves	289	379	368
Pigs	466	1,046	799
Sows	--	293	852
Rams	49	--	111
Ewes	858	1,227	865
Lambs	478	--	464
Horses	171	99	152
Hens	--	3,296	--
Geese	300	300	--
Turkeys	--	90	--
Ducks	--	12	--

The standard of animal husbandry practice on the island was like that practised by the Acadians on the mainland and was not very high. The livestock depended mainly on what they could procure by roaming in the woods and vacant lands. The limited areas of marsh provided some winter fodder, but the amounts harvested usually were skimpy for the number of animals maintained. The pigs depended on the mast from the trees in the forest. One description stated that they roamed in the woods in the summer and fattened on the beechnuts in the fall.[21]

The extent of livestock holdings varied considerably and we can get a picture of this from the details in the census of 1752, taken by Sieur de la Roque. A few settlers were listed as having no livestock, though most had some. Of those with minimum holdings were: one pig and one hen; one pig and six fowl. Some of the larger holdings were: 6 oxen, 4 bulls, 9 cows, 6 calves, 1 horse, 1 sow, 3 pigs, 5 sheep; 8 oxen, 2 bulls, 6 cows, 1 heifer, 3 calves, 2 horses, 5 ewes, 3 sows, 3 pigs; 5 oxen, 3 bulls, 5 cows, 2 heifers, 1 calf, 8 wethers, 14 ewes, 4 sows, 6 pigs. The largest holding was: 8 oxen, 8 cows, 8 calves, 1 horse, 1 mare, 30 ewes or wethers, 17 pigs, 21 geese, 11 turkeys, 12 fowl. The owner of this holding also possessed a flour mill. Most of the settlers had at least two oxen which would indicate that they had the motive power required for tillage.[22]

It is interesting to note that this particular census was recorded in narrative rather than in tabular form. A few examples of entries are: "They have in livestock five oxen, four cows, one sow and four pigs. The land on which they are settled is situated on the Riviere des Blancs, it has been given verbally by M. de Bonnaventure. They have made a clearing on it of five arpents in extent where they have sown seven bushels of wheat and eight bushels of oats. In livestock they have one pig, having lost all besides during the winter. Of livestock they have two oxen, one cow, one horse, one wether, two ewes, two sows, four pigs, five geese and ten fowls. The land on which they are settled is held in grant from M. Duvivier. They have made a clearing for the sowing of thirty-two bushels, but have sown only seven not having been able to procure more owing to their poverty."[23]

The horse population was relatively small but it was of concern to the government officials. Prevost[24] suggested that, in view of the fact that horses consumed more forage than cattle, they should be restricted in number to not more than one per family. He thought that this could assist in increasing beef production for export to Louisburg.

The available evidence indicates quite clearly that the crops on Prince Edward Island, during the French regime, were subjected to a series of natural disasters, apparently not experienced to the same extent by the Acadians in Nova Scotia and New Brunswick. One of the more devastating occurrences was the plague of mice that appeared from time to time. The first one was in 1724 and was followed by another in 1728. That year De Brouillon[25] reported that a favourable crop was in prospect when the mice appeared and destroyed the grain as well as the grass. The destruction was so complete that the farmers plowed down the residues. The crop loss left them in a precarious position for food and seed for the next year and made it necessary to purchase seed from Acadia for the next crop.

Lenormant[26] refers to the complete destruction of crops by mice in 1738. Roma, at Brudenel, experienced this latter plague and wrote a treatise on mice in which he contended that this pest would disappear as settlement increased and the forests were removed. He stated that the mice propagated in the forests and advanced from there to destroy the crops.[27] Sieur de la Roque described the animals as follows:

"These animals resemble in appearance those found in the rural districts of France, especially in Champagne, where during the fall, they store up, at a depth of two or three feet in the earth, grain for their comfortable subsistence and go to sleep for six months of the year. It is only in this foresight that the field mice of this country do not resemble those of the old land, for here, after they have devoured everything that they can find to their taste in the country, they throw themselves into the water where they are drowned in such prodigious numbers that their bodies form a kind of dam to the waters, by which they are carried down and accumulate, so that the shores of the lakes, rivers, creeks, and streams are filled with them..."[28]

Another hazard, directly related to upland agriculture, was the forest fires that occurred from time to time and destroyed not only the forests but the crops on the cleared land, and also destroyed buildings and livestock. One of these was in 1736 when a fire swept through the most thickly settled district and destroyed the crops.[29] Again, in 1742, a fire swept through the St. Peters area in the month of June and destroyed crops, property, and took the lives of 13 people.[30]

Plant diseases also made their appearance and in 1740 rust damaged the wheat crop so extensively that a severe shortage in the harvest was the result.[31] In 1751 there was a report that: "the wheat crop was totally scalded."[32]

Insect problems were not unknown and in 1749 grasshoppers appeared and ruined the crops.[33] Also in 1750 there is a description of another attack: "The second bad season was caused by innumerable legions of locusts of a prodigious size. They were so voracious a species that they ravaged all the growing grain, vegetables and even the grass and buds on the trees."[34]

Absentee Owners

With the fall of Louisburg the French lost Isle Saint Jean which now became island St. John under English rule, with the subsequent deportation and emigration of the major part of the French population. No exact figure is available, but it was estimated by Villejouin[35] that the population was reduced to about 250 habitants. In other words an almost completely new start had to be made. Developments under English rule came slowly, partly because of the manner in which land grants were made. A complete land survey of the island was made by Holland, beginning in 1765. This divided the island into 67 townships or lots. One lot was reserved as crown land and the other lots were granted to individuals, most of whom had very little, if any, real interest in developing their property. In fact, what turned out to be largely absentee proprietorship was a handicap to development for over a hundred years.

A description of conditions on the island at that time is given by Holland: "The ground is in general covered with strawberries in their different seasons which are very good; with proper care, it produces most kinds of grain, wheat, barley, oats, peas, beans, etc.; also cabbage, cauliflower and potatoes, very good, in great abundance; carrots, turnips, etc. In those places which have been settled, and are still tolerably cleared, is very good grass; but a great part of the lands formerly cleared are so overgrown with brush and small wood, that it will be extremely difficult to form a true estimate of the cleared lands, or to make it fit for the plough again. It may be proper to observe here, that very few houses mentioned in the explanation of the Townships are good for anything, and by no means tenantable, except one or two at St. Peters, kept in repair by the officers, and one built by me at Observation Cove."[36]

Shortly after his arrival in 1770, Patterson, the first governor of the island as a separate colony, provided his evaluation of the situation: "The soil appears to be very good, and easily cultivated. It is reddish in colour, mixed with Sand; and in most places free from stone....I never saw finer Grass in my Life, than grows on every place where it is clear of Woods. It will produce every kind of Grain and Vegetables common in England with little or no trouble; and such as I have seen of the latter are much better of their kinds than those at home; tho raised in a very slovenly manner."

"The Bears in some parts destroy the Sheep; and the Mice this Year are so plenty that they have in most places destroyed the little Grain which was attempted to be raised... The Inhabitants say their appearing in such numbers is periodical once every seven years."

"The French inhabitants have for some years past been mostly maintained by a few British Subjects here, who have employed them during the summer in fishery, and have been paid their wages in Cloathes, Rum, Flower, Powder and Shot, with the last articles they kill as many Bears, Seals, and wild Fowls as serve them for meat, the Seal Oyl they call their butter and use it as such. By this means Agriculture has been so much neglected, there is not one Bushell of Corn raised by all the French Inhabitants on the Island."

"There are a few British Families who have raised some. There have arrived here this summer about one hundred and Twenty Families part sent by Mr. Montgomery the Lord Advocate of Scotland, the rest by a Mr. Stewart of that country."[37]

Immigration from Britain continued at a relatively slow rate for many years and was augmented by the arrival of Loyalists after the American revolution. By 1805 the population total was only about 7,000. One group of settlers of particular interest was brought over by Lord Selkirk in 1803. A description of their experience is of interest as it may well picture what all of the early settlers faced on arrival: "Buying out several of the island's proprietors, Selkirk sent some 800 people, mostly from the Isle of Skye, to Orwell Bay in 1803. In this region perhaps 500 to 1,000 Acadians had resided briefly fifty years before, but it had remained almost completely unoccupied since. It compares well in quality with most of the island's lands, but Selkirk, certainly unaware of this, must have found its chief attractiveness its emptiness and relative cheapness."

"The settlers, who arrived in early August, were for the most part very poor with little capital even in the essentials of tools, stock and seeds. Selkirk secured a minimum of supplies to see them over the first winter, but the prospect was not promising for the crofters with no experience of the transatlantic climate and forests. There was evidence in wells and shallow ditches of the French attempt to establish a foothold, but the heavily overgrown "clearings" were as forbidding as the unaltered forest. The almost miraculous speed with which they learned the axe skills of the North American pioneer saved their situation. In a few weeks they had made adequate shelters for the winter and throughout that season continued to clear land for the first season's crops. The annual routine of extending the cleared land and practising husbandry of oats and potatoes changed little for years, but they did convert the temporary shelters of the first winter to substantial houses rather rapidly."[38]

An insight into the important part that potatoes played for early settlers, and also an indication of their hardships, is found in Curtis' description of his experiences during the winter that he spent on the island in 1775-76. "They gave us the best they had to Eat and Drink being Salt Ells and Potatoes but the poor Creatures they had but little or nothing but water to Drink."[39] "I soon found my friend Compton & wife who were so recovered from their fatiege by rest, but very bad off for provisions having nothing but Salt Codfish and Potatoes with plenty of good water."[40] "Having nothing now to eat but Salt fish and potatoes for diner we use to have hot and other times cold this kind of food three times a day."[41]

The general level of husbandry practised by the Anglo-Saxons was no better than that of the earlier French settlers. In this regard the situation on the island was similar to that in Nova Scotia and New Brunswick at the same time, the underlying reasons being the same in all cases. A good picture of the situation is provided by the letters of Johnstone, who spent the years 1820-22 travelling throughout the island. He gave a general description of the layout of the settlements and then commented on agricultural management:

"Allow me now to take a more particular view of their farms, their stock, crop, and method of management. Their farms are generally 100 acres, English measure, half a quarter of a mile broad, and a mile and a quarter long. They are laid out narrow, to get as many farms upon the shores or sides of rivers as possible. This will be found in the end an inconvenient form; but in the outset it does well enough, as every new settler is anxious to have a piece of his front wholly cleared as early as possible, in order to see his neighbour's house, and to be near enough to visit him occasionally. And perhaps the first generation, according to the progress many of the settlers have made, will not clear more than 20 or 30 acres all their life, and the distant end of the farm is carefully kept for firewood for future generations. And thus rent is paid for lands yearly, from which no particular advantage will be derived perhaps for sixty years to come. But their want of steady industrious habits, particularly among the youth born upon the Island; stout horses or oxen to work with; good implements of agriculture; lime easily procured; knowledge how to make and preserve dung; good roads, and remunerating prices for agricultural produce, added to the delight they take in fishing and eating fish; their general poverty and ignorance in managing a clear farm — all these, and many other things, contribute to make them bad farmers; and where they have large clear farms, they are letting much of the land run entirely wild and barren. They know not how to drill potatoes and turnips, nor do they even seem to consider that ploughing these frequently between the furrows, both kills weeds and enriches the soil."[42]

Johnstone referred to the lack of care and use of dung again and again in his letters and apparently thought that this was a major weakness in the farm operations.

Another observer, some five years later, also commented on the rather poor practices being followed: "The general mode of conducting a Farm is slovenly, often wretched. Cattle, sheep and pigs are turned into the woods, or on the shore, to get their own living during Summer; and, frequently, as much time is lost in seeking the stock, as would clear enough land to support them in good pasture. Few farms have any subdivision fences. A patch is ploughed here for wheat, another there for barley, the intervening spots are mown for hay; and yet, under all this want of judicious arrangement, it is astonishing what returns are obtained — a like management in England would not give the farmer bread and cheese. But there are many meritorious exceptions to this cobbling sort of system. Many farmers display in their management an accurate and intimate knowledge of their difficult calling; these are developing the powers of the Island soil; and their example, in connection with the exertions of the Agricultural Society, are operating a great and satisfactory change."[43]

The Trail Blazers of Canadian Agriculture

This observer could see some progress being made. Whether this was the result of the writings of Agricola, who had had considerable effect on the agriculture of Nova Scotia and New Brunswick at this same time, cannot be determined but forces were at work that would continue to have a beneficial effect.

As with the Acadians and the French population on the island, livestock played a major role with the new settlers, but at the time of Johnstone's visit the system of management still was at a low level. He made frequent references to their stock and his descriptions give a good picture of the general situation:

"Their horses are rather small, and light in make, but uncommonly hardy and spirited. Their black cattle are all horned, and some of them are stunted in appearance; but it is no wonder, for they are fed upon wheat straw the greater part of the winter, and often allowed only a scanty portion of the same. When they have to drive them early in the summer to the woods in their weak condition, they are in danger of getting mired in the swampy places, and in this way several are lost. They are obliged to keep all the calves sucking at home, to entice the cows to return from the woods at night, but they have to wander so far before the cravings of nature are satisfied, that even this inducement fails to draw them home sometimes for a night or two together. When this happens, their milk is greatly injured, or altogether lost. To remedy this they hang a bell to the neck of the one they have most dependence upon, and if she leads the way home, the rest will generally follow. This bell serves also another valuable purpose, namely, to find out their retreat in the woods, when the people are obliged to go in search of them themselves. Their sheep are of the whitefaced kind mostly, but lean and long legged. They are exceedingly healthy, and produce fine wool, though in small quantities. They are never laid with tar, and have to lie in the house all winter, and to be fed with upland hay; and as their cattle run at large in the woods during the summer, they almost all of them keep more stock than they have winter feed for, consequently all their fodder is eaten up, and their dunghills are not half so big as might be formed from the dung of the same number of cattle in Scotland."[44]

Further he stated:"They have no green feeding for their cattle in winter, for they never think of giving a service of potatoes; and the swine are so poorly fed, that if they get hold of a fowl they will eat it alive. I was even told of a man who had a weak cow eaten to death at the stake by the pigs. From the poor way in which their cattle are fed during the winter, some of them die of weakness, or when driven out to the woods in this state, they are more in danger of getting mired, as well as of falling prey to the wild beasts. This, and the great travel they have in the woods, cause them to give very little milk, although it is said to yield a good return of butter. Their sheep, though very healthy and prolific, sometimes fall a prey to foxes and wild cats; as to the number they keep,

I cannot speak with anything like certainty. New settlers, unless they get marsh hay along with their farms at first, get slowly on in keeping stock."[45]

Not only did the settlers have mammals but: "They have turkies, geese, ducks and common poultry, the same as in Britain, but they must all be kept close in the house over winter; and their out houses are generally so open and cold, that if they do lay eggs, they are generally rent with the frost before they are noticed. Some of the farmers keep very large flocks of every kind."[46]

The equipment available to the farmers continued to be relatively primitive and Johnstone commented: "Their implements of agriculture are very deficient indeed; except a few ploughs that have been imported from Britain, they have not many of their own making that deserve the name. They use a very broad sock or share in the form of the Lothian plough, and sometimes lock the coulter and sock together by making a hole in the backside of the former, in which the point of the latter is inserted. This is for the purpose of ploughing rooty land. They make very broad furrows, and I apprehend not deep enough, and lay them too flat...They run almost all their cart wheels without any iron round them, the soil being so free from stones or gravel, that bare timber lasts a long time. But they haul, as they call it, every thing in the winter upon sledges or slayes, as they name them, upon which, when they have good snow or ice to go upon, they carry a great load."[47]

Only five years later another observer noted a favourable change in this respect. "There are good ploughs in the Colony, and all implements, harness, houses, etc. are rapidly improving; but there is still a miserable deficiency of winnowing machines, chaff-cutters, and other labour saving implements. A good machine for washing clothes is a desiratum."[48]

The primitive nature of the operations was described rather vividly by Johnstone:

"After the wood is all burnt, the stumps are left standing about two feet high, scorched black with the first burning, like so many blocks of a blacksmith's anvil. The people then begin planting their potatoes, which is done in the following manner: with their hoes they scratch or rake a little of the earth to one side, about eight inches square, and after raking a little of the ashes lying upon the surface into this groove, they place four cuttings of seed potatoes in it in square form, and then cover them up with earth till it resembles a small mole-hill, and still repeating the same operation they go on putting all their seed into the ground by four at a time, and when the space cleared is all planted it looks as if it were all covered with small mole-hills. And this is the only labour bestowed upon the potatoes till they are ready for raising. But when the time for planting arrives, man, wife, children, and all that can handle a hoe, must work, as the season is short; and if the crop is not got in to a sufficient extent, want may stare them in the face, when a supply will be difficult to procure, and when there will be nothing in their pocket to pay for it. This work of planting with the hoes is very laborious, for there are always a great number of small roots spread over the surface, which they have to cut to pieces with their hoes, otherwise they could not plant it at all. They are also much hurried by reason of the wood, which will not burn early in the season, and which, as well as the land, is rendered extremely damp by the melting of the snow. But there is one thing much in their favour; the heat that the fire leaves in the ground is so great, or some other cause supervenes, to hurry vegetation so rapidly, that they can plant potatoes a great deal later on their new burnt land than on any other, and still have a good crop. I have been well assured of their planting upon this kind of land as late as the 12th of July, and after all have little to complain of. I may observe, that when the fire runs well the first time it is put to the wood, not only a great deal of leaves and rubbish are burnt, but also a part of the surface soil. This enriches the land greatly, and their first crop, as well as the succeeding one, are, in this case very luxuriant. I have been credibly informed, that the increase has been in favourable seasons, from twenty-five to upwards of thirty-fold; but the average may be taken from fifteen to twenty. To proceed, however, after a crop of potatoes has been taken, next year the same ground is sown with wheat and timothy grass seed. This crop is generally hoed in or harrowed with a harrow in the form of the letter A, with the point foremost; after this, if the land is a good quality, and has been well burnt, they can mow it several years among the stumps, but generally it will not bear to be mowed till the stumps are sufficiently rotten for stumping... When only a part of the stumps are got out, a kind of ploughing is made among them the preceding fall, and next spring it is ploughed again and sown with oats; after the oats are reaped, the remaining stumps are taken out, and it is ploughed in the fall, and next spring potatoes set with dung upon it; and the following year it is sown with wheat and timothy grass seeds, and laid down for upland hay, as they call it; then it is mown year after year, as I said before, for perhaps eight or nine years without any additional manure whatever. This is the whole story of clearing land upon the island."[49]

Livestock Improvement

One of the first official actions to encourage improved agriculture was taken in 1823 when "His Excellency the Lieutenant-Governor, (in Council) has been pleased to direct the sum of Ten Pounds to be distributed in premiums, for the best of the following articles, which shall be produced at the New Market House in Charlotte-Town in the first day of its opening.

Prince Edward Island

	Cattle		£ s p
For the best Carcass of Ox Beef			2 0 0
do	do	Cow beef	1 0 0
do	do	Pork	1 0 0
do	do	Mutton	1 0 0
	Grain		£ s p
For the best sample of Wheat not less than a Sack of 3 Bushels			1 0 0
do	do	Barley	1 0 0
do	do	Oats	1 0 0
do	do	Field Peas	1 0 0
do	do	Horse or Tick Beans	1 0 0"[50]

That some action was being taken by individuals to improve the livestock situation can be seen from a letter to the editor August 30, 1823:

"While our grain stands unrivalled, our dairy cattle are much inferior to those of Nova Scotia. And this disadvantage appears more evident as our farms expand, but our breed of sheep reflect anything but credit, on our farmers. Our fleeces do not on an average, weigh two pounds and a half each, whilst the sheep lately imported by Messrs. Stewart, Chanter, and Goff, from the United Kingdom weigh upwards of nine pounds each. Few of us have the opportunity of making such importations from Europe, but there are plenty of good sheep on the Continent. We have already found that a good pig can be fattened more easily than a bad one, and the breed of that valuable animal has been considerably improved here lately. Mr. Sims killed one which weighed 740 lbs."[51]

This letter provides the first record of direct importation of stock. But numerous additional importations were made in the following years. The pattern of multiplicity of breeds so characteristic of present-day Canadian livestock production had an early origin as we see from a listing of stock imported in the period 1826-1830 as recorded by Lewellin:[52]

Stock Imported
1826
Colonel Ready, late Lieut. Governor
    Thorough-bred Horse Roncesvalles
    do   do   Mare Roulette
    Alderney Bull and Cow
    Northallerton Polled Cow
    A Southdown Ram; Six Southdown Ewes
    One Ram, Southdown and Leicester crossed
    A Leicester Ram; A Berkshire Sow
    A Suffolk Bull Calf and two Berkshire Pigs died on the passage.

1830
    A Suffolk, polled (purchased by the Agricultural Society).
    A Southdown Ram, Bred by Elman of Glynd, in Sussex.

Mr. Braddock - 1826
    Ram and Ewe, Leicester crossed with long Scotch, Messrs.L. & A. Cambridge
    Glamorgan Bull and Heifer.
    A Devonshire Bull and Cow.

A Suffolk Cow, (now at Mr. Duck's farm, Winter River).
The Devonshire Bull "Tommy Chanter", now the property of the Agricultural Society.
A Devonshire Bull, now in the possession of Mr. Haviland.
A Devonshire Cow, now in the possession of Mr. Worrell.
Leicester Rams. Leicester Ewes.

Mr. Lewellin
　Half-bred Suffolk entire,　do　Mare.
　A long-horned Northern Cow, famous for milk and meat, and breeding fine stock.
　A Sow descended from a Chinese Sow and Monmouthshire Boar, and afterwards repeatedly crossed with good Berkshire Boars.
　Two Rams and three Ewes, from the Cotswold Hills, Gloucestershire, a present from Mr. Sulivan.

Hon. G. Wright
　One Southdown and one Leicester Ram, 1829.

Mr. Billing
　A Devonshire Heifer.

Mr. Stone
　The half-bred Horse "Wanton", since exported, has left a few foals behind him.

Mr. Tremaln
　A Boar and two Sows, of the pure Chinese breed.

Mr. Benbow
　A Ram and some Ewes of the long wooled Lincoln breed, also some pigs, the breed I forget.

Mr. Worthy
　A Ram, Leicester and Southdown crossed, and a Leicester Ewe.

Charles Stewart, Esq. of Princetown Royalty, also imported, some years ago, a Ram from England, which has greatly improved the breed of sheep in his neighbourhood.

In the above listing reference is made to the Agricultural Society. This society came into being in February 1827 with the stated purpose "that the object of this Society shall be, to diffuse and extend the knowledge and promote the practice of the best and most approved modes of Agriculture. To encourage the breed of horses, cattle, sheep, and swine, by importation of new Stock, and by the judicious crossing of the old. To promote the making of the most improved implements of husbandry. To encourage both the growth and importation of seed grain, of all sorts of grass seed, and the seed of roots. To encourage the clearance of forest land, and to promote a better mode of cultivating it than now prevails, to encourage the growth of wool and dressing of cloth."[53]

The society lost no time in getting into action and made plans for an exhibition the same fall. The prize list was quite modest compared to those of today:

	£	s	p
"For the best Ram	2	0	0
second best　do	1	0	0
For the best Heifer	1	10	0
second best　do	1	0	0
For the best Ewe	1	0	0
second best　do		10	0
For the best Wedder	1	10	0
second best　do	1	0	0

For the best Pig	1 10 0
second best do	1 0 0
For the best bag of Wheat, 3 bushels	2 0 0
" " " " Barley do	2 0 0
" " " " Oats do	2 0 0
" " " " Tick Beans 1 bushel	1 0 0
" " " " Carrots do	1 0 0

The above Premiums to be paid to the actual Feeders and Growers only."[54]

The report[55] of the results of the exhibition is interesting as it shows the active participation of government officials as exhibitors and also provides some descriptions of the animals shown. Clearly the standards of that time were very different from those of today.

The Agricultural Society originally was limited to Charlottetown but as early as December 1827 moves were made to establish another one in the Bedeque settlement.[56] In the 1829 annual report of the Charlottetown Society reference is made to the fact that no public funds had yet been made available to the society. However, that soon changed for later that year the Legislative Assembly made a grant of £150. The society continued to function and new societies were formed, all receiving financial support from the government. For example, in 1834, £120 were granted with "Forty Pounds to be paid to the Central Agricultural Society for the purchase of Seeds — Twenty Pounds to each of the Agricultural Societies in Prince County, provided the number not exceed two; and Twenty Pounds to each of the Agricultural Societies in King County, provided the number does not exceed two — the grants to said societies to be expended in the purchase of Seeds."[57]

The number of societies increased to eleven by 1840. They arranged for livestock and grain shows, ploughing matches, and other activities in addition to importing livestock and seeds. In many cases the Central Society acted as agent for the district societies in importations. An example of importations by the Society is found in the 1829 annual report:[58]

```
40 bus. seed Barley        10 bus.  Talavera Wheat
20 "  Poland Oats          10 "     Winter   do
20 "  Black Oats           2 Cwt.   Dutch Clover
8 "  White Field Peas      2 "      Cow Grass
4 "  Large Horse Beans     50 Lb.   Mangel Wurtzel
4 "  Small   do   do       A drill machine
```

Other items imported are shown in an 1837 invoice:[59]

```
31 Bars best hammered Iron at 3d per lb.
50 Wilkies Plough mounting, light size 7s. 6d.
100 do      do         middle size 9s.
6 Cart Axle Blocks and Bushes
12 Malleable Iron made Socks and Heads 13s.
35 yds. No 10 16 inch Iron Wire Web 4s. 6d.
20 Fanner Wheels and Pinions 5s. 6d.
5 Bus. Grey Field Peas 8s.
6 Cart Saddles, Girths and Breeches
2 Neck Collars
```

2 Sacks fine Dutch White Clover seed
11 lbs. at 1s. 3d. per lb.

The items imported usually were sold at cost to members of the societies and at a slightly higher price to non-members. A complaint voiced at one of the meetings was that some members sometimes bought items and resold them to non-member friends or neighbours at cost. This tended to inhibit these non-members from becoming members.

In 1840 the Central Society reported the purchase of a ready made fanner, a horse rake, a cultivator, a chaff cutter, and several other small implements. "The Fanner, a very superior set, and the Horse Rake, which was put in operation in a field, and performed satisfactorily, are kept as patterns at the Society's depot, where they can at any time be seen. The increase of Threshing Machines will give the farmer a choice of these labour saving implements. Ten pounds have lately been awarded to Mr. William Smallwood, of Township Forty-eight, for improvements on the Horse Power, capable of driving a Threshing Mill."[60] One of the threshing machines referred to was of local manufacture, first advertised in 1837, by a Mr. Bovyer, as follows: "These machines will be warranted to thresh 80 to 100 bushels of grain per day, with one horse, and two persons to attend same."[61]

In 1841 the Society could report: "Your Committee are highly gratified that the demand for labour-saving machines is very much on the increase among farmers. Mr. Stephen Bovyer's Threshing Machine is much prized, and is coming into pretty general use."[62] At the same time they reported on other progress stating that before the advent of the agricultural societies "the clumsy wooden plough was universal — now there are very few districts in which one is to be met with. Upwards of three thousand sets of iron plough mountings have been imported and disposed of by the Society."[63] In 1843 the Society reported that 33 of Bovyer's threshing machines were in operation.

The societies also were active in encouraging other aspects of improvement in agriculture, one being the introduction of lime as a soil additive. For example: "The Eastern Agricultural Society are desirous of engaging a Person who understands the burning of Lime for Agricultural purposes, and who will superintend the manufacture of about 80 to 100 Tons of Stone into Lime, during the ensuing Summer — the Kiln to be at Mr. George Aitken's Lot 59."[64] Some progress was made in the use of lime for in 1841 the Society reported that "they are happy to find that Lime is at length coming into use, and from some of the experiments that have recently been made in the Royalty of Charlottetown, especially on the farms of Mr. George Beer, sen. and Mr. John M'Gill, it is evident, that the productiveness of our soil is capable of being vastly increased by the judicious application of lime."[65]

The societies also undertook or supported some simple experiments. For example, in 1840 a new variety of potato, the Rohan, was tested against the Scotch Apple, the Blue, and the White varieties.[66] A little later some guano was imported for test and an attempt was made to utilize guano from some of the islands in the St. Lawrence.

Ploughing matches were conducted in various parts of the province and the report on one, in 1837, provides a picture of the cultural practices in use. "The Rules prescribed by the Society were — each furrow slice to be ploughed 4 1/2 inches deep, and eight inches wide — to plough all alike, commence in the centre, and gather the ridge — that the cattle should not be hurried, as goodness of work was the object — to adopt the ridge furrow slice, as in broadcasting seed, it naturally falls into the lowest part of the furrow, and then by harrowing, the seed is deeper covered, and comes up in rows, giving a better chance for free circulation of air in its after growth, and to clean by hoeing, and probably some saving of seed — having nearly the same effect as sowing by drill machine."[67]

While the above examples indicate some progress resulting from the efforts of the societies all was not well. Thus, a letter to the editor, regarding the cattle show of 1837 stated: "The bulls were wretched. Where are the descendants of the famed Tommy Chanter? Very few superior animals appeared on the grounds. The vaunted breed of Dishley or New Leicester sheep do not appear to hold their own: they appear to be too much a park stock to flourish on beech leaves."[68] This indicates, among other things, that the nutrition being supplied to livestock had not yet reached the desired level.

The Eastern Society also commented on the cattle situation in 1840: "It has been a matter of general remark that, in this section of the Island, neat Cattle have much fallen off in size, few oxen would feed to more than 500 lbs. This may be owing, in a great measure, to a largely multiplied Stock; and a decreased extent of pasturage outside the Fences; and it suggests the absolute need of subdividing every Farm by permanent, and, if practicable, line Fences, and feeding nearly the whole of the Stock on the land."[69]

Importations of livestock continued intermittently by the Society as well as by individuals. Thus, in 1837, the Lieutenant Governor imported "two Guernsey Cows, an improvement on the much esteemed Alderney; a thorough bred Mare, stinted to Bolero; a very fine Leicester Ram; seven Ewes, and several Pigs."[70] The same year the Society agreed to import: "two 2-year old Bulls and two 2-year old Heifers of the Ayrshire breed from Scotland. An entire horse from Canada — aged from three to five years. A 3yr old entire horse, Clydesdale, from Scotland. Also, some pigs of the Berkshire breed."[71]

The government continued to give financial support to these activities but at times placed definite stipulations on the manner in which the grants were to be used. For example, in 1845, a special grant of 150 pounds was allocated to the Central Society for the purchase and importation of a horse. It was stipulated that the horse was to be sold at auction on arrival and the purchaser was to agree to have it stand for service, for the first two years, three days every month in each of the three counties, for the months of May, June, and July. It was also stipulated that if the sale price was sufficiently great the grant funds should be returned to the government.[72]

Despite incomplete success in improving agricultural practices and the livestock population there is little doubt that the societies were an important factor in bringing about desirable changes. There were problems for which the societies could provide no relief, namely, problems of weather, insects, weeds, and crop diseases. The hazards of production encountered by the early French settler and the ones that came later, continued to hamper development. Patterson[73] made reference to the mouse plague in 1770, but there appear to be no later reports of this pest as a destructive force. But crop diseases continued to cause partial or complete crop failures from time to time. Smut, rust, and mildew were the common diseases of grain and especially of wheat. Rust was reported quite regularly in wheat and occasionally in other grains. In 1845 the Agricultural Society report emphasized the importance of this problem and suggested one possible means of overcoming it:

"The wheat, which promised an unusually abundant return, has been severely affected by its great enemy, the rust. The variety called Tea Wheat, having, in many situations, escaped this evil, recommends itself to the farming community for general cultivation, even although it might be, in productiveness, inferior to some other kinds. Oats would have been very plentiful, but they too have been greatly affected with rust and blight — a most unusual circumstance."[74]

Potato diseases first came into prominence in the 1830s. In 1838 the Eastern Agricultural Society commented: "Your Committee remark with pleasure, that although disease has prevailed to a considerable extent, and in some cases proved fatal, yet we have been so far favoured by a bountiful Providence, during the past season, as to remove all apprehension of scarcity. Among other mercies, we have to record, that in Scotland the evil which affected and threatened the total destruction of the potato, has been removed, and in this Colony also in a great measure — let us hope that the injury may totally cease next season. Nevertheless, it would be prudent to renew the potato by seedlings, which should be encouraged by premiums offered by the Central Society."[75]

Apparently the disease did not cause major concern until 1845 but then it struck in full force as can be seen from the reports in the Register for September 16 and 23: "We have also been informed by a friend who has just returned from a tour to the southward, that the Potato crops, which but a short time ago promised such abundance, will in many places be a total failure, in consequence of being affected by rot; — the stalks appear to decay prematurely near the ground, and disease ascends to the top, causing a black curl in the leaf, and is not as may be supposed, confined to wet soils. The first appearance of the disease is under the skin, of a spotted brownish colour; it runs through the potato, which when boiled, emits a disagreeable smell, quite different from common sourness, and if dug up for a day or two, they will be quite rotten."[76]

"The present season, is the first within the memory of the oldest inhabitant, in which such a disease as that which now appears to prevail, has been known to affect that valuable esculent, although Canada and New England suffered in the same way last year. It appears however during the present season not to be confined to the American Continent, for in England, Ireland, Belgium and some other European States, a similar failure is reported, although whether arising from the same cause, or resulting in the same manner, we have not sufficiently clear accounts to guide to a correct conclusion."[77] The Island apparently was caught up in the general problem that affected other parts of Canada and other parts of the world.

Johnstone gives a vivid description of some of the insect problems: "Another evil also followed; insects of various tribes were brought to life and maturity, as if by a miracle; the blood thirsty mosquitoes sprung up in thousands, and attacking the face, the hands, and legs, if these are only covered with stockings, they manifest neither fear nor shame in prosecuting their designs of sucking blood, and instilling their poison. Grasshoppers next made their appearance, some of them rattling their wings in the air, and others leaping before one so thick upon the road that it seemed all in motion. In a still evening they would join in millions, in singing a kind of chirley song like a flock of birds at home, in winter.... To defend the cattle from the flies in the night, they kindle a large fire in their fold or pen, and the black cattle and sheep will contend who to get nearest the smoke till their hair and wool are sometimes singed."[78]

In 1845 another threat appeared with reports of a new insect attacking the grain in the stack.[79] The first report apparently was somewhat exaggerated, but there was a problem though the identity of the insect is not clear. In 1854 there was a report of the weevil that attacked the grain in the ear.[80] This possibly was the wheat midge which was a serious pest in New Brunswick at this same time.

It is not clear how serious a problem weeds were as they are not mentioned in the society reports. Johnstone described one: "There is a small weed which they call yar (spirie) which greatly damages the crops; on some lands, when once it gets into the ground, there is no method yet discovered of clearing it away. When the seed ripens, it will lie under ground, I have heard, for twenty years, and vegetate as fresh as ever when the land is again plowed up. I have seen wheat damaged so much by it, as not to be more than a fourth of an average crop."[81]

Hazards of weather also caused losses at times. In 1837 the Assembly found it necessary to pass an act to provide seed grain and potatoes to certain settlers because an early frost in 1836 had destroyed their crops. In 1844 hail storms were reported to have damaged crops severely in various localities.

Legislation Enacted

Reference has been made to certain cases of official action in support of agriculture. Additional indications of official concern are the acts passed from time to time to regulate or support the industry. Some of the early acts of interest are:[82]

1773
- An act for indemnifying persons who shall burn the small bushes, rotten windfalls, decayed leaves and all other brush and rubbish upon the Lands and in the Woods upon this Island. The reason for this act, as outlined in the preamble, is interesting. "Whereas there is great reason to believe that the burning of the small bushes, rotten windfalls, decayed leaves, and all other brush and rubbish upon the Lands and in the Woods, tend to the necessary clearing and cultivating of the same, and the production of Grass and Herbage for depasturing cattle, as well as to the destruction of Insects some of which are troublesome to the labouring people." The burning was limited to April, May, and June.

1780
- An act to prevent the cutting of Pine or other trees without permission of the proprietor, and to prevent the cutting down of Fences.
- An act to prevent the Running at large of Stallions or Stone horses and the killing of Partridges at Improper Seasons.
- An act for the prevention of Trespasses by unruly Horses, Cattle, Hogs and Sheep and for preventing the running of Hogs at large through the Town of Charlotte Town.

1781
- An act giving a Reward for the killing of Bears.

1790
- An act to prevent the malicious killing, wounding, and maiming of cattle. This referred to horses, neat cattle, swine, and sheep.
- An act for compelling persons owning lands adjoining to each other, to make their respective parts or proportions of the fence between them, and for empowering the Grand Jury to nominate fence viewers. This act defined a lawful fence as being four and one half feet high.
- An act to prevent the running at large of rams at improper seasons.

1795
- An act to prevent the running at large of Geese within the Town of Charlotte Town.
- An act for ascertaining the Standard of Weights and measures in this Island.

1796
- An act to prevent the robbing of gardens and orchards, potatoes, and turnip fields, and throwing down fences.
- An act for the preservation of sheep, throughout the island. This imposed a tax on dogs.
- An act for appointing commissioners of sewers. This referred to dykes and drains in the marshes and not to sewers as the term is used today.

1801
- An act for granting a bounty for growing and cultivating hemp in this island, for the purpose of exportation.

1825
- An act to establish a reward for the killing of bears and loupcerviers.
- An act to prevent the running at large of boar pigs, and to restrain swine from going at large without rings.
- An act to prevent the destruction of sheep by dogs.

1828
- An act for establishing the standard weight of grain and pulse, and for appointing proper officers for measuring and weighing the same.

Wheat shall weigh		59 pounds	58 pounds
Rye	do	57 do	56 do
Indian corn	do	58 do	57 do
Barley	do	49 do	48 do
Oats	do	40 do	36 do
Pease	do	60 do	
Beans	do	60 do	

1829
- The above act was amended to reduce the weights as shown in the right hand column above. It had been found that the locally grown grain could not measure up to the original weights.

1831
- An act to encourage the Settlement and Improvement of Lands on this Island, and to regulate the proceedings of a Court of Escheats thereon.

1837
- An act to provide Seed Grain and Potatoes for certain Settlers and to regulate the distribution and mode of repayment.
- An act to prohibit the exportation of Grain, Meal, and Potatoes and for other purposes therein mentioned. This was for one year.

1844
- An act for raising a Fund for the encouragement of Agriculture to be expended in the erection of Lime Kilns and the burning of lime. This imposed a new tax of one farthing per acre on cultivated land — one half penny on uncultivated land.

The Trail Blazers of Canadian Agriculture

In concluding this chapter the statistical data in Table II provide a picture of some of the important changes that have taken place in Prince Edward Island agriculture from mid-19th century to recent times. Unlike the situation in the other three Atlantic provinces, agriculture in Prince Edward Island has increased, though certain crops have declined in importance.

Table II
Population and agricultural statistics for selected years

	1734[a]	1753[b]	1834[c]	1861[d]	1966[e]
Population	572	2,256	33,292	80,857	108,535
Improved land/acres	--	--	387,616	368,127	569,799
Horses	--	152	6,299	18,765	4,967
Cattle	332	2,317	30,428	60,012	125,224
Sheep	119	1,440	50,510	107,245	15,106
Pigs	--	1,651	20,702	38,553	82,907
Hens/chickens	--	3,296*	--	--	352,461
Wheat/bu	--	--	128,350	346,125	97,000
Barley/bu	--	--	38,850	223,195	433,000
Oats/bu	--	--	261,664	2,218,578	3,519,000
Buckwheat/bu	--	--	--	50,127	--
Potatoes/bu	--	--	1,310,063	2,972,333	7,341,000
Tame hay/tons	--	--	--	21,088	246,000
Butter/lbs	--	--	--	711,487	5,967,000

* Data for 1752
a. Data from NA MG1 G1 466/43
b. Data from NA MG1 G1 466/47
c. Data from NA C0229/8 Journal of the House of Assembly Appendix C p275
d. Data from *Census of Canada* 1668-1871.Ottawa, Taylor, I.B. 1876 Vol.V.
e. Data under this heading are from the 1966 census for the first seven items. The other items are for 1965 from the 1967 *Canada Year Book*.

Chapter V
Quebec

The first agricultural activities by Europeans in Quebec were carried out by Cartier and his men during his third trip to Canada in 1541. This was a planned colonizing effort comprising five vessels and a large number of people. "Enough food was brought to last two years. Besides agricultural tools and ready-made carts, a number of domestic animals were put on board to work on the land and 'to multiply there': 20 cows, 4 bulls, 100 sheep, 100 goats, 10 hogs and 20 horses or mares. These were to be the first European livestock in the St. Lawrence Valley, if not in all of North America north of Florida."[1]

Cartier chose to establish the settlement above the present site of the city of Quebec rather than at the site where he had spent the winter on his previous trip. At Cap Rouge, where he arrived in August 1541, great activity ensued. Some of the people turned to constructing the necessary buildings to house the people and the livestock while others set about preparing land for cropping despite the lateness of the season. As Trudel writes: "While some worked at building, the others worked the land; in one day, twenty men ploughed an acre and a half and sowed the seeds of 'Cabbages, Naveaus or small Turnips, Lettises, and others, which grew and sprong up out of the ground in eight dayes'."[2] It may be questioned that we should take literally the term "ploughed" to mean the use of a plough. The fact that 20 men were involved would suggest that handwork, with hoes or other tools, was involved. Be that as it may, it is clear that a considerable area of land was prepared for cropping.

Despite all of the preparations that had been made, the experience of the winter was such that Cartier left in June 1542 with all of the settlers. There is no mention of what happened to the livestock that had been brought though it is unlikely that any were left in this country. Shortly after Cartier's departure, Roberval, who was to have accompanied Cartier in the first instance, arrived with a new complement of settlers, supplies, and livestock. About all that he found of the previous occupation was some wheat that Cartier had sown before his departure. Roberval and his settlers did not fare any better than Cartier with the result that the settlement was abandoned in 1543.

More than sixty years were to pass before there again was agricultural activity in Quebec. In 1608, the team of De Monts and Champlain, who earlier had been together in the Acadia settlement, decided to establish a trading post at Quebec as a means of getting ready access to the furs of Canada. Champlain was given command of the expedition, a single ship with a cargo of things necessary for the venture. That some agricultural activity was planned is indicated by the fact that a gardener was included among the men. He had a contract to go to Canada for two years. Confirmation of such planning is provided in Champlain's report: "Whilst the carpenters, sawyers and other workmen were busy at our quarters, I set all the rest to work clearing the land about our settlement in order to make gardens in which to sow grains and seed, for the purpose of seeing how the whole thing would succeed, particularly since the soil seemed very good."[3] Later he wrote: "On the first of October I had some wheat sown, and on the fifteenth some rye. On the third of the month there was a white frost and on the fifteenth the leaves of the trees began to fall. On the twenty-fourth of the month I had some native vines planted, and they prospered extremely well, but after I left the settlement to come to France, they were all ruined, for want of care, which distressed me very much."[4]

It has been reported that in this year he planted some apple trees from Normandy and these survived the Canadian climate so well that 25 years later they were still yielding. Champlain does not record this directly but in 1616 he noted: "There was also some very fine Indian corn, and grafts, and trees which we had brought thither."[5]

There is no detailed record of what transpired agriculturally the first year or two but some success must have been achieved for, in 1610, Champlain returned to France and reported: "As for the gardens, we left these well provided with kitchen vegetables of all sorts, as also with very fine Indian corn, with wheat, rye and barley, which had been sown, and with vines which I had planted there during my winter's stay."[6]

Though Champlain did undertake some agricultural activity, Louis Hebert is usually credited with being the first farmer in Quebec. Hebert first went to Port Royal where he indicated an interest in husbandry. He came to Quebec in 1617 under contract to the company that had the trading monopoly at that time. He was accompanied by his wife and three children and his brother-in-law. Under his contract he was to do whatever might be asked of him and he was to care for the sick free of charge. With the permission of the agent he might work on the land that he had been given, seven acres, in his leisure time but the results of this work would belong to the company. At the end of his two-year contract he would be freed from his obligations, but even then he would be permitted to sell his produce only to the company.[7] Despite the limitations of time and equipment, he cleared some land and soon possessed a small patch of crop and in this way he set an example for others to follow, though not many did.

Information regarding the time of first arrival of various classes of livestock is fragmentary and sometimes contradictory. There is only indirect evidence that pigs were included in the supplies of 1608, but Champlain reported that the carrion of a sow and a dog were thrown out during the winter to attract foxes.[8] Letourneau[9] stated that the first horned cattle were debarked at Quebec in 1610, but that the first pigs appeared in 1634. This is in contradiction to Trudel's[10] statement that, in 1620, the Recollets did not have cows and goats but did have pigs, geese, and poultry, as well as a male and female donkey. We can be quite sure that there were cattle at Quebec before 1623 because that year Champlain wrote: "Recognizing the inconvenience we had suffered in past years through making hay so late for the cattle, I had two thousand bundles made up at Cap Tourmente as early as the month of August, and sent one of our pinnaces to bring them up."[11] Sagard corroborated the presence of cattle in 1623, at the time of his visit, when he wrote: "There is another dwelling on top of the height, in a very convenient spot where a number of cattle brought from France are pastured."[12]

It is evident that in the early years of settlement the amount of land under cultivation was limited. In 1618 Champlain noted: "I inspected everything, the cultivated land which I found sown and filled with fine grain, the gardens full of all kinds of plants, such as cabbages, radishes, lettuce, purslain, sorrel, parsley, and other plants, squash, cucumbers, melons, peas, beans and other vegetables as fine and as well forward as in France, together with vines, brought and planted here, already well advanced, in short everything increasing and growing visibly; not that the praise, after God, must be given either to the labourers or the manure that was placed there, for as may be believed, there is not much of it, but to the excellence and richness of the soil, which in itself is naturally good and fertile in all kinds of advantages as experience has shown, and one might get increase and profit from it, both by tilling and cultivating it and by planting fruit trees and vines, as also in feeding cattle and French barnyard fowls."[13]

But the fact remains that the settlement was far from self-sufficient as far as food was concerned and depended on annual shipments from France. Thus, in 1629, when the English were besieging Quebec, and Champlain was faced with the prospect of feeding the population without the usual supplies from France, he wrote: "Awaiting the harvest of peas and other grains on the cleared lands of the widow Hebert and her son-in-law, who had between six and seven arpents sown, that being the only recourse I had:... As to the Reverend Jesuit Fathers, they had only enough cleared land in seed to serve for themselves and their serving men to the number of twelve...The Recollet Fathers had a good deal more land cleared and sown, and they were but four in number. They promised us that if they had more than they required from their four or five arpents of land sown with several sorts of grain, vegetables, roots, and garden herbs, they would give us some."[14] This description would indicate that even after 25 years of development less than 15 acres of land were under cultivation.

Two factors were responsible for this slow development. The first was

The Trail Blazers of Canadian Agriculture

the failure of the monopoly company to give any real support to colonization, being primarily interested in the fur trade. The second reason was that only the most primitive tools were available for cultivation. All of the early cultivation was done by hand tools, the pickaxe, spade, and hoe. As Champlain noted: "...until an inhabitant of the country should seek the means of relieving the men who usually worked with their hands in cultivating the soil; which was not broken by the ploughshare drawn by oxen, until the twenty-fourth of April 1628."[15]

While the main effort at development was at Quebec, as early as 1611, Champlain made the first tentative moves at what was later to become Montreal. As he reported: "Now whilst waiting for the Indians, I had two gardens made, one in the meadows and the other in the woods, which I caused to be cleared. And on the second day of June I sowed some seeds there which all came up quickly and in perfect condition, which shows the good quality of the soil."[16] Nothing further was done at this site until many years later.

The first land concession in the St. Lawrence valley was granted on February 4, 1623, to none other than Louis Hebert when he was granted, in perpetuity, the land that he occupied on the cape.[17]

Another important development was at Cap Tormente where de Caen had property. In 1623 he visited this area with Champlain and decided that the cape, with its broad meadows, would be a suitable place for stock raising. It was decided to develop this as a centre, but it was not until 1626 that action was taken. At that time a stable, sixty feet long by twenty wide, and some dwellings were constructed.[18]

The settlers struggled against great odds, including neglect by the sponsoring group. But worse was yet to come in the form of an attack by the English under David Kirke. His men raided and destroyed the Cap Tormente establishment in 1628, and in 1629 took possession of Quebec, which was not returned to French control until 1632. During the intervening years no agricultural development took place, the English being content to use the land already under cultivation.

With the return of Canada to the French, new developments were undertaken. An insight into the agricultural activities can be found in the reports of Father Le Jeune: "The men are the horses and oxen; they carry or drag the wood, trees, or stones; they till the soil, they harrow... We have two fat sows which are each suckling four little pigs, and these have been obliged to feed all summer in our open court. Father Masse has raised these animals for us. If the point of which I have spoken were enclosed, they could be put there and during the summer nothing need be given them to eat; I mean that in a short time we shall be provided with pork, an article that would save us 400 livres. As to butter, we have two cows, two little heifers, and a little bull. M. de Caen having left his cattle here when he saw he was ruined, we took of them three cows, and, for the family which is here, three others. They and we gave M. Gifford a cow so we have remaining the number I have just stated. For lack of a building, they cost us more than they are worth, for our working people are obliged to neglect necessary things for them; they spoil what we have sown; and they cannot be herded in the woods, for the insects torment them. They have come three years too soon, but they would have died if we had not taken them in; we took them when they were running wild. In time they will provide butter, and the oxen can be used for plowing, and will occasionally furnish meat. As to grains, some people are inclined to think that the land where we are is too cold. Let us proceed systematically, and consider the nature of the soil; these last two years all the vegetables, which came up only too fast, have been eaten by insects, which come either from the neighborhood of the woods, or from the land which has not yet been worked or purified, nor exposed to the air. In midsummer these insects die, and we have fine vegetables."

"As to fruit trees, I do not know how they will turn out. We have two double rows of them, one of a hundred feet or more, the other larger, planted on either side with wild trees which are well rooted. We have eight or ten rows of apple and pear trees, which are also well rooted; we shall see how they will succeed. I have an idea that the cold is very injurious to the fruit, but in a few years we shall know from experience. Formerly, some fine apples were seen here."

"As to Indian corn, it ripened very nicely the past year, but this year it is not so fine. As to peas, I have seen no good ones here; their growth is too rapid. They succeed very well with this family, who live in a higher more airy location."

"The rye has succeeded very well for two years. We planted some as an experiment, and it is very fine. Barley succeeds also. There remains wheat; we sowed some in the autumn at different times; in some places it was lost under the snow, in others it was so preserved that no finer wheat can be seen in France. We do not yet know very well which time it is best to take before winter to put in the seed; the family living here has always sown spring wheat, which ripens nicely in their soil. We sowed a little of it this year, and will see whether it ripens. So these are the qualities of our soil."[19]

From the foregoing it is evident that a learning process was under way. The settlers had come into a virtually unknown environment, different from that of their native land and only by experimenting were they able to develop a system of agriculture suitable to the new territory.

A few years later Le Jeune recounted more of their experiences and from this it is clear that they had learned from the various trials that they had made.

"March wheat sown in the spring succeeds better than wheat sown before winter. Not that I have not seen some fine wheat that was sown in October. But as we are not yet thoroughly acquainted with the weather and the nature of the soil and climate, it is safer to sow in the spring than before winter. Common barley and hulled barley succeed to perfection, and rye does well; at least I can assert that I have seen all of these grains grow as beautiful as they have in France. The peas are better and more tender than those they bring over in ships. Pot-herbs do very well, but the seeds must be brought over. It is true that the nearness of the forest, and so much rotten wood, of which the land is, as it were, formed and nourished, engender, at times, insects which gnaw everything, as these animals die during the heat of summer, everything comes to perfection, but sometimes later than is desirable to secure the grain and seed. We have oxen and cows, which we use to cultivate the cleared; this year some asses have been brought over, which will be of great service; horses could be used, but there is no hurry about bringing them."[20]

Natural hazards continued to give trouble. In his report of 1633 Le Jeune wrote: "The woods are troublesome; they retain the cold, engender the slight frosts, and produce great quantities of vermin, such as grasshoppers, worms, and insects which are especially destructive in our garden; we shall rid ourselves of them, little by little, without, however, leaving this place."[21] A few years later he again reported on caterpillars and grasshoppers so it is evident that they continued to be a problem.[22]

He also referred to the weather: "The heat here is intense and burning, and yet I have observed since I have been here that there has been frost every month of the year...I have never experienced in France anything like the heat and drought which we have had here during June. Everything in the earth burns in such weather, and yet it froze one morning in the house of the Recollet Fathers."[23]

In 1642-43 he reported: "Notwithstanding all these hindrances, nearly every French household now provides its little store of wheat, rye, peas, barley, and other grains necessary to the life of man — some more, some less — some making provision for nearly six months, others for only a part of that time. Now they begin to understand the nature of the place and the right seasons for tilling the soil."[24]

Policies Discouraged Farming

From this it is clear that, after 35 years, the settlers still were far from self-sufficient in food. The various citations from Le Jeune's writings indicate that they had learned much about the crops that could be grown successfully, but also that there were natural obstacles to easy success. But these were not the main reasons for lack of achieving self-sufficiency. The reasons are to be found in the policies of the officials responsible for the colony. Trading with the Indians for furs continued to be the main interest to the practical exclusion of all else. Development of an expanding agriculture was the least of their concerns.

In these early years, settlement continued to be concentrated at Quebec City, with a few settlers at Trois Rivieres, Sillery, and Beauport. In 1642, a small settlement was established at Montreal, with 55 people staying on through the first winter. Agricultural activity was begun at that site that year. Eccles notes: "enough land was cleared to plant a little wheat and peas, and to graze the two oxen, three cows, and twenty sheep that had survived the voyage."[25]

De Casson differed slightly as he stated: "from the time this habitation was begun some peas and Indian corn had been sown, and this was attended each year."[26] Referring to 1644 he stated: "This sharp fight and several others which took place during the year did not prevent people this spring from sowing French wheat at the instigation of M. d'Aillebout, to whom Canada owes this first trial, by which everyone was convinced that the coldness of this climate did not prevent the growing of a great deal of grain."[27] In this he ignores the fact that this had already been well demonstrated by the people at Quebec. However, grain did succeed at Montreal and the first mill was established there in 1648.[28]

An indication of the development of the Montreal colony is given by Eccles: "With thirty arpents of cleared land an habitant could provide enough for his family. By 1662 three quarters of the settlers produced enough for their needs and many families had a surplus. Though not yet quite self-sufficient in the essential foodstuffs, it was clear that the colony soon would be."[29]

It may be noted here that, while cattle, sheep, goats, and pigs were brought over with the early settlers, it was not until 1647 that the first horse, since Cartier's abortive settlement, was brought in. There is no record of what happened to this animal, but no additional horses were imported until 1665.

While the fur trade was the primary interest of the association responsible for the early development, some limited efforts at land settlement were made. One important element in this was the seignorial system. Under this system, large grants of land were made to favoured individuals who then were expected to bring in settlers for development of the grants. In many cases these conditions were not fulfilled. The way the system worked is described by Lanctot:

"The seignor received his concession from the proprietary company. He then conceded dependent seignories in occasional cases, but regularly granted lands *en censive*. The censitaries, or tenant farmers, at once set about clearing the land. Cutting down the tall trees of the forest was slow and exhausting work. Each man could prepare only an acre and a half a year for the plough. For his food the hired man was given every week two loaves of bread of about six pounds, that is, about a barrel of flour a year. Along with this went two pounds of lard, two ounces of butter, a small measure of oil and vinegar, a pound of dried codfish and a pint of peas. Their drink consisted of jar of cider or beer daily and a small amount of wine on holy days. In winter, the men preferred 'a bit of brandy in the morning'. Game and fish which were abundant and close at hand completed their diet. The life of the freeholder was a sober industrious one, but the soil was rich and rewarded him the following year with a plentiful harvest of wheat, vegetables and fruits."[30]

The primary interest of the individuals and associations who held concessions continued to be in the fur trade and colonization never did receive strong, active support. As a result, it was not until the crown took over direct responsibility for the colony in 1659 that serious steps were taken to increase settlement and develop the country. An early step was to send a strong military force to subdue the Iroquois. When this was accomplished discharged soldiers were urged to accept grants of land in the colony. Some success was achieved from this for: "Once the war was over and the menace of Iroquois attacks ceased, agriculture made rapid advances. New land was wrested from the forests, particularly around Montreal, as discharged soldiers took up concessions and the sons of habitants, who had remained on their fathers' farms for security reasons, began to clear land of their own. The decline of the fur trade also served as a stimulus to agriculture."[31]

Other action was taken and, in the period of substantial subsidization of agricultural colonization which ended in 1673, a total of 4,000 new inhabitants had been brought to the colony. Thus, the population increased from a mere 200 or 300 in 1640 to 2,500 in 1663, 3,215 in 1665, and 6,705 in 1673. Included in this increase were the women who were shipped over to become wives of the male settlers. An example of an individual case was; "After the choice was determined, the marriage was concluded on the spot, in the presence of a Priest, and a publick Notary; and the next day the Governor-General bestowed upon the married Couple, a Bull, a Cow, a Hog, a Sow, a Cock, a Hen, two Barrels of salt Meat, and eleven Crowns."[32]

Demonstration Farms

During this period some livestock was imported, beginning with a shipment of 14 horses in 1665 and a total of 55 before 1670. Forty five sheep were sent in 1667 and 44 in 1668. There apparently already were sufficient swine in the colony and Talon, the Intendant, saw no need for bringing any more. The problem lay in preventing those already there from becoming a nuisance by damaging and destroying crops and, by 1673, the Council was establishing regulations for their control.

Talon gave stimulation and encouragement in several ways. He possibly was the first to recognize the need for demonstration farms. In new villages he reserved lots for established settlers capable of teaching the newcomers how best to operate. On his own property on the St. Charles river he maintained horses and a fine herd of cattle. He raised all sorts of poultry for distribution among the habitants. He was also interested in introducing new plants and planted hops on his farm and then built a brewery to provide a market for barley as well as to provide beer for the settlers. In 1666, he planted hemp and distributed hemp seed in an effort to establish this crop. In encouraging the settlers to raise hemp he used rather strong-armed methods by seizing all the yarn in the colony and giving it back only to those who promised to repay in hemp. When some did produce the crop he bought it in order to induce others to produce it. Despite his efforts, and those of his successors, including the English administrators, it never was possible to develop a strong desire for this crop among the habitants.[33]

In 1671 Talon wrote to the King and suggested that if there was no objection, he would encourage the settlers to cultivate tobacco. The King did object, partly because he thought that this might interfere with the production of wheat, but also because tobacco was an important crop in other royal domains and he did not want competition. Despite this, tobacco production did occur, but it did not appear in colonial statistics until 1721. The actual beginning of production is not known but, in view of the reported production of 48,038 pounds in 1721, it is evident that production had been under way for a considerable number of years prior to that time. Tobacco production was quite general, in a small way, on many, if not most, of the farms, but mainly for home consumption.[34]

A picture of agricultural development during the latter third of the 17th century and the first third of the 18th century can be formed from the data in Table I. Not only does this show the growth of population, cultivated land, and production, but it indicates, in a general way, when the various items of statistical data first were recorded.

The data in Table I show that the livestock population had a fairly steady growth and there was increase in the production of wheat, oats, peas, flax, and tobacco. Hemp never became a popular crop despite the encouragement given to its production. The habitants preferred to raise flax because it was easier to handle, especially in the processing procedure. Furthermore, flax provided them with material they could use whereas hemp was primarily a marketable crop. Indian corn never became an important crop, the climate of the Quebec City area particularly being unfavorable for its growth.

Table I
Population and agricultural statistics for the period 1667 to 1735

		1667	1683	1712	1721	1735
Population	No.	4,871	10,251	18,732	24,511	37,271
Cultivated land	Arpents	11,174	25,217	52,965	62,145	164,741
Wheat	Minots	--	--	297,415	282,700	669,744
Corn	"	--	--	6,353	7,205	5,215
Barley	"	--	--	--	4,585	3,446
Oats	"	--	--	57,191	64,035	152,681
Peas	"	--	--	45,893	57,400	70,848
Flax	Lbs.	--	--	52,098	54,650	92,489
Hemp	"	--	--	5,795	2,100	9,590
Tobacco	"	--	--	--	48,038	195,340
Cattle	No.	2,136	7,025	17,630	23,388	36,870
Sheep	"	--	625	6,575	13,823	20,576
Swine	"	--	2,330	10,095	16,250	24,962
Horses	"	--	56	3,208	5,603	8,381

Data are from NA MG1 G1-460 and G1-461.

The increase in horse population was of concern to the authorities and efforts were made to limit it. In 1710, the King ordered that what he considered to be an overplus of horses in the colony should be caused to die out by isolating the mares and castrating the stallions.[35] The concern had two facets. The one was that horse production interfered with the production of cattle, the latter being considered to be of greater importance to the colony. The second point was completely different as can be seen from a memorial of 1712: "The settlers have no need of them except to till their land and haul their wood and their grain. It is not natural for the settlers to use them in winter to travel from place to place, instead of going on snowshoes, as they should all do."[36] The next year the King again emphasized the need for carrying out his instructions to reduce the number of horses as he stated: "It is important that the settlers should be made to return to the use of snowshoes, and the horses and winter vehicles be destroyed, otherwise they would become effeminate and lose their superiority."[37] Despite all this the horse population continued to increase at a rate even greater than the population increase.

Like the Acadians, the French in Quebec did not have the potato until very late in the French regime. According to Salone: "The potato was cultivated in Canada before the English conquest. In 1758 Vaudreuil and Bigot received orders to make it known in the colony and they were able to respond that this already was the case, that a habitant had brought some from France, and he had found that it yielded very well with very little effort."[38] Shortly before this Kalm had reported: "Few people took any notice of potatoes, and neither the common nor the Bermuda ones were planted in Canada; only a few had artichokes. When the French here are asked why they do not plant potatoes, they answer that they do not like them and they laugh at the English who are so fond of them."[39]

During the latter part of the 17th century the colony produced enough grain in normal years to meet its own requirements. In good years there was some for export but when poor crops occurred, as they did with some regularity, export restrictions were imposed and imports had to be made from France. Even at this early stage of development there were attempts at supply management by government. Eccles[40] relates that because of a large wheat surplus, the Sovereign Council, in 1664, purchased a thousand bushels, at the generous price of five livres a bushel. In order to encourage the habitants to grow more grain. The difficulty encountered in supply management arose in part from the fact that the wheat was not usually threshed until mid-winter and thus the amount available for the next year could not be estimated closely before the last ships sailed for France in November. In order to ensure that the wheat supply would be adequate for the troops and the civilian population, the intendant was likely to over-estimate the amount of flour that would be required from France for the next year. "When there was a sizable surplus in the colony, as in 1694, he had to have it exported to the West Indies and to France to prevent the price in the colony subsiding too much and the habitants planting less the following year as a result."[41]

In years of poor crops real hardship, even famine conditions, could occur. Severe crop losses in 1737, 1739, and 1742 led to such conditions. The situation in 1738 was described as follows: "I can't express to you, monsieur, the misery that was caused by the shortage or famine that was felt in the whole country. A very large number of the inhabitants, especially on the south shore needed bread for a long time and a large number had gone to the north shore, which was less maltreated, to receive charity and some grain to seed. The others lived on a little oats and Indian corn and fish. I have been in constant necessity to furnish the indigents, the infirm, and the poor regularly with bread, meat, and vegetables. M. Michel has done the same at Montreal on my orders. Otherwise we would have exposed whole families of subjects of the king to death."[42]

We come now to the end of the regime of France in Quebec. Unlike the situation in Acadia, the Quebec population was permitted to remain and continued relatively intact. This meant that new immigrants had to settle on uncleared land except in cases where they could buy cleared land from previous owners. Uncleared land could be bought from the seignors or could be obtained as a grant from the government. Immigration was limited mainly to the merchant class between the time of the conquest, in 1759, and the influx of Loyalists in the early 1780s.

Before proceeding to developments after the conquest we should review the general agricultural situation at that time. The best description of the agricultural practices is provided by Kalm, a Swedish natu-

The Trail Blazers of Canadian Agriculture

ralist, who visited the colony in 1750. He came by way of New York and described conditions in three different areas. At his entry into the colony at Fort St. John the following:

"The harrows which they made use of here are made entirely of wood of a triangular form. The plows seem to be less serviceable. The wheels upon which the plow beam is placed, are as thick as the wheels of a cart, and all the wood work is so clumsily made that it requires a horse to draw the plow over the surface of the field."[43]

"About four French miles from Fort St. John, the country had quite another appearance. It was all cultivated, and a continual variety of fields with excellent wheat, peas, and oats, presented itself to our view; but we saw no other kinds of grain. The farms stood scattered, and each of them was surrounded by its own corn fields and meadows... All the fields were covered with grain, and they generally used summer wheat here. The ground is still very fertile, so that there is no occasion for leaving it fallow."[44]

At Laprairie: "The grain fields around this place are extensive and sown with summer wheat, but rye, barley, and corn are never seen."[45]

At Quebec: "Kitchen herbs succeed very well here. White cabbage is very fine but sometimes suffers from worms. Onions are very much in use here together with other species of leeks. They likewise plant several species of pumpkins, melons, lettuce, wild chicory or wild endive, several kinds of peas, beans, Turkish beans, carrots and cucumbers. They have plenty of red beets, horseradish, and common radishes, thyme, and marjoram. Turnips are sown in abundance and used chiefly in winter. Parsnips are sometimes eaten, though not very commonly. Few people took notice of potatoes, and neither the common nor the Bermuda ones were planted in Canada; only a few had artichokes...Those who have been employed in sowing and planting kitchen herbs in Canada and have had some experience in gardening told me that they were obliged to send for fresh seeds from France every year, because they commonly lose their strength here in the third generation and do not produce such plants as would equal the original ones in quality."[46]

"The scythes are like our Swedish ones; the men mow and the women rake. The hay was prepared in much the same way as with us, but the tools are a little different. The head of the rake is smaller, has tines on both sides and is a little heavier. The hay is raked into rows; and they also use a kind of wooden fork for both pitching and raking. In so doing, however, a good deal of hay is left on the field since this does not rake as clean as an ordinary rake... The hay is taken away in four-wheeled carts drawn by either horses or oxen. The oxen are hitched in such a way as to pull with their horns instead of their shoulders... The grass lots are usually without fences, the cattle being in pastures on the other side of the woods and cowherds take care of them where they are necessary."

"The grain fields are pretty large. I saw no ditches anywhere, though they seemed to be needed in some places. They are divided into ridges, of the breadth of two or three yards broad, between the shallow furrows. The perpendicular height of the middle of the ridge, from the level of the ground is near one foot. All the grain is summer sown, for as the cold in winter destroys the grain which lies in the ground, it is never sown in autumn. I found white wheat most common in the fields. There are likewise large fields with peas, oats, in some places summer rye, and now and then barley. Near almost every farm I found cabbages, pumpkins, and melons. The fields are not always sown, but lie fallow every two years. The fallow fields not being plowed in summer the weeds grow without restraint in them and the cattle are allowed to roam over them all season. There was a superabundance of fences around here, since every farm was isolated and the fields divided into small pastures."[47]

"All the horses in Canada are strong, well-built, swift, as tall as the horses of our cavalry, and of a breed imported from France. The inhabitants have the custom of docking the tails of their horses, which is rather severe treatment, as they cannot defend themselves against the numerous gnats, gadflies, and horseflies. They put the horses in tandem before their carts, which has probably occasioned the docking of their tails, as the horses would hurt the eyes of those behind them, by moving their tails to and fro... It is a general complaint that the country people are beginning to keep too many horses, by which means the cows are kept short of food in winter."

"The cows likewise have been imported from France, and are of the size of our common Swedish cows. Everybody agreed that the cattle, which are born of the original French breed, never grow to the same size as the parent stock. This they ascribe to the cold winters, during which they are obliged to put their cattle into stables and give them but little food. Almost all the cows have horns; a few, however, I have seen without them."

"Every countryman commonly keeps a few sheep which supply him with as much wool as he needs to clothe himself... The sheep degenerate here after they are imported from France, their wool becoming coarser, and their progeny is still poorer. The want of food in winter is said to cause this degeneration. I have not seen any goats in Canada, and I have been assured that there are none."[48]

Under date of August 29th he wrote: "The harvest was now at hand, and I saw all the people at work in the grain fields. They had begun to reap wheat and oats a week ago."[49]

"Every countryman sows as much flax as he wants for his own use. They had already harvested it some time ago, and spread it on the fields, meadows, and pastures, in order to bleach it."[50]

"Every farmer plants a quantity of tobacco near his house, in propor-

The Trail Blazers of Canadian Agriculture

tion to the size of his family. It is necessary that one should plant tobacco, because it is so universally smoked by the common people."[51]

By September 19th Kalm was back at Montreal and wrote: "Several people here in town have gotten French grapevines and planted them in their gardens. They have two kinds of grapes, one of a pale green, or almost white; the other, of a reddish brown... The cold of winter obliges them to put dung round the roots of the vines, without which they would be killed by the frost."[52]

"One half of the grain fields are left fallow, alternately. The fallow grounds are never plowed in summer, so the cattle can feed upon the weeds that grow on them. All the seed used here is spring seed, as I have before observed. Some plow the fallow grounds late in autumn, others defer that business till spring; but the first way is said to give the better crop. Wheat, barley and oats are harrowed, but peas are plowed into the ground. Farmers sow commonly about the fifteenth of April, and begin with peas. Among the many kinds of peas which are to be gotten here, they prefer the green ones to all others for sowing. Peas require high, dry, poor soil, mixed with coarse sand. (They did not know what it meant to stake peas.) The harvest time commences about the end and sometimes in the middle of August. Wheat returns are generally fifteenfold and sometimes twentyfold; oats from fifteenfold to thirtyfold. The crop of peas sometimes fortyfold, but at other times only tenfold, for it varies very much. The plow and harrow are the only implements of husbandry they have and those are not the best sort. The manure is spread out in spring. They sow no more barley than is necessary for the cattle, for they make no malt here. They sow a good deal of oats, but merely for the horses and other cattle."

"I have been assured that some people who live on the Isle of Jesus sow wheat in autumn, which is better, finer, and gives a more plentiful crop than spring wheat; but it does not ripen more than a week before other wheat."

"Some kinds of fruit trees succeed very well near Montreal, and I had here an opportunity of seeing very fine pears and apples of various sorts. Near Quebec the pear trees will not grow because the winter is too severe for them, and sometimes they are killed by frost in the neighbourhood of Montreal. Plum trees of several sorts which were first brought over from France, succeed very well and withstand the rigors of winter."[53]

Kalm's report indicates that, in the 100 years or so between his writing and that of Le Jeune, the settlers had learned to adapt crop production to the different conditions of the various parts of the province. Kalm's report leaves the impression of a relatively successful and secure system of agriculture. However, this is somewhat at variance with the situation described by General Murray, in 1762, after the English occupation:

"With very slight cultivation, all sorts of grain are here easily produced, in great abundance, the inhabitants are inclinable enough to be lazy, and not much skilled in Husbandry, the great dependancies they have hitherto had on the Gun and fishing rod, made them neglect tillage beyond the requisites of their own consumption and the few purchases they needed, the Monopolies that were carried on here in every branch, made them careless of acquiring beyond the present use, and their being often sent on distant parties and detachments, to serve the particular purpose of greedy and avaricious men without the least view to public utility, were circumstances under which no country could thrive. As they will not be subject to such inconvenience under a British Government, and being necessarily deprived of arms they must of course apply more closely to the culture of their lands."[54]

In this Murray may have been showing some bias but the substance of his statement is supported by numerous French-Canadian writers. For example, Ouellet[55] states clearly that the level of agricultural practices at the time of the conquest was at a low level and the general state of agriculture was very bad. He ascribes much of the blame to mismanagement by and corruption in the colonial government which led to inflation and low morale. However, a basic cause was the lack of knowledge, among the farmers, of improved cultural practices and this had led to the gradual depletion of the original fertility of the soils to the point where yields were declining.

Unfortunately, no major changes in farming practices among the French-Canadian farmers occurred following the conquest and the level of production continued to be very low, though some exceptions did occur. A limiting factor was the seignorial system which, through the years had failed to encourage agriculture; in fact it had tended to stifle its development. This became increasingly evident after the English occupation when the seigneurs, in an effort to increase their revenues, imposed increased rentals on their tenants, a move that became self-defeating as the impoverishment of the soil led to continual decrease in productivity. A contributing factor was the population increase without a proportionate increase in land clearing; cleared land was subdivided to provide for farmers' sons.

Some of the policies of the French colonial government were continued. For example, with regard to hemp production Murray wrote: "They grow both Hemp and Flax in some parts of the Country, and many of the lands are well cultivated for this production. It will be right to turn the thoughts of the people towards the cultivation of this article, so essential to Great Britain and for which she annually pays great sums to Foreigners, a few premiums properly disposed of, some Germans and Russians skilled in raising and preparing the same and encouraged for that purpose to become settlers here may in a short time greatly improve this most useful branch of agriculture."

"This will be one means of employing the women and children during the long winters in breaking and preparing the Flax and Hemp for exportation, will divert them from manufacturing coarse things for their own use, as it will enable them to purchase, those of a better sort manufactured and imported from Britain."[56]

It is clear from Murray's statement that Canada now was looked upon as a supply base for certain essential products and as a market for British goods.

Even the policy with regard to horses did not change as can be seen from the following: "As probably it may be thought right not to receive the duties on dry goods, a Tax upon Horses might be introduced in lieu thereof, it would serve also to restrain a piece of luxury the people of this country are too apt to run into, in that respect, and prove a means to encourage the breed of horned Cattle of which at present, by the unavoidable waste of War, they are very short, besides as Cattle must be housed here for a long time during the winter, the Horned kind are foddered with more ease, less cost, and afford more utility."[57]

While there was a considerable influx of the English merchant class into the province following the occupation, the number of farmer immigrants was limited until the Loyalists arrived after the American Revolution. Most of those who came during the early period settled in and around Quebec City and Montreal where they, in many cases, purchased land that had been farmed for many years to the point of soil exhaustion. One group of Loyalists settled in the Gaspé area but a greater number from the states of Vermont, New Hampshire, and New York moved across the border into what became known as the Eastern Townships. For a long time they continued to have closer ties with the states from which they came than they had with the settlers at Quebec and Montreal because of the poor means of communication with the latter. Even after roads were developed so that movement between the areas was facilitated the distances involved dictated in part the system of agriculture followed. Livestock production became more profitable than grain production because the cost of moving it was relatively less.

The coming of the English-speaking farmers was seen to offer some hope for improvement in agricultural practices. Thus Campbell, who visited in 1791-92, wrote: "In the Island of Montreal, of eighteen miles long, are many spacious and fine farms, some of which are possessed by Englishmen who cultivate and manure their land as is done in that country, and raise crops which astonish the natives, who now begin to follow their example, and will soon, it is to be hoped, spread over all the populous and extensive province of Lower Canada."[58]

Weld, reporting two or three years later, wrote: "The style of farming amongst the generality of the French Canadians has hitherto been very slovenly; manure has been but rarely used, the earth just lightly turned up with the plough, and without any other preparation the grain sown;

more than half of the fields also have been left without any fences whatsoever, exposed to the ravages of cattle. The people are beginning now, however, to be more industrious and better farmers, owing to the increased demand for grain for exportation, and to the advice and encouragement given to them by the English merchants at Quebec and Montreal, who send agents through the country to the farmers to buy up all the corn they can spare."[59]

The statement by Weld indicates that the arrival of the English merchants marked the beginning of a change of attitude with regard to the role of agriculture. While fur continued to be an important element in trade, the new merchants also recognized the important role that agriculture could play and set about encouraging improved practices to increase production.

A hopeful stimulus to the improvement of agriculture was the formation of the Quebec Agricultural Society in 1789, the first general meeting being held on April 6 of that year. A similar society was formed at Montreal on November 2 of the same year. Interest in the Quebec Society was apparently great and enthusiastic with 83 subscribers at the first meeting. In the course of the next few years this interest waned and, in 1793, the annual meeting had to be rescheduled as an insufficient number of members showed up at the appointed time to provide a quorum. The same thing happened the following year.[60]

Early Research Efforts

It is pertinent to note some of the activities of the Society as this indicates the problems that seemed to be of major concern and shows clearly the serious limitations of the knowledge available at the time. Among the resolutions passed at the first meeting was: "That the society make the means of preventing smut from affecting the crops of wheat and oats one of the first objects of their enquiry, and that the best method of preparing seed be recommended for trial the ensuing season throughout the province." As a result, the directors, on December 21, 1789, "ordered that the process in preparing seed wheat as communicated by Mr. Cartier, be published in the Quebec Gazette and the Herald. From the experiment made by Monsr. J. Cartier in the Parish of St. Antoine on the Chambly River, laid before the directors of this Branch, it appears that smut in wheat, may be effectively prevented by using the following precaution."

"Let the wheat intended for seed be thrice thoroughly washed, and let the water drain from it after the third washing; Then steep it eighteen hours in brine strong enough to float an egg, and spread it on the floor to let the brine run off, but whilst the wheat is yet moist, let quick lime be equally sifted over it, stirring it very well with a shovel, and continue sifting on more lime until the wheat is equally dusted with it, in the proportion of half a gallon for every bushel of wheat. By stirring it well with the shovel the wheat will soon be dry and fit for sowing."

"Some exceeding smutty wheat or what the inhabitants call 'bled charbonne' was prepared by Mr. Cartier in the manner above directed and immediately sown; the sample sent to the Directors of this Branch, as the produce of that very smutty wheat, was good, clean, well filled grain."[61]

On March 11, 1790, an additional procedure was published by the Society and is given here to show that they were anxious to try any procedure that offered some possibility of controlling smut. "Preparation recommended by Mr. Couillard: To clean 240 pounds of wheat, requires 25 quarts of water, 2 1/2 pounds of dung of poultry, the same quantity of sheep dung, or instead thereof so much of that of pidgeons. The mixture must steep twelve or fifteen days in a tub, and be stirred now and then with a stick; at the expiration of which term it must be drained clear from the sediment. Take a part of the liquor and warm it; dissolve in it three pounds of slack lime, or a pound and a half of quick lime. If at the time of dissolution the effervescence should be too strong, it must be checked by means of a small quantity of cold water. This lime water must then be mixed with the overplus of the infusion of dung, then put the 240 pounds of wheat (after it has been washed, scummed and dried properly) in that liquor to steep the space of 10 minutes. Then spread it and stir it often till the next day, when it is fit for sowing."

"Mr. Couillard has likewise successfully used human urine and chimney soot for the preparation, in which case he diminishes the doses of the other components. He assures that in repeated trials the above modes of preparation have been attended with much success; nothing being more effectual in preventing smut. He says the produce of wheat prepared after this manner is at the rate of four sheaves to a bushel and 1/4 and about 2 1/6 fine wheat, without a grain of smut."[62]

In the light of modern knowledge, the above methods appear to be very crude, and possibly not very effective. Further evidence of the Society's efforts at experimenting and testing, as well as in solving other problems, is found in the president's address to the 1790 annual meeting. "His Excellency the Governor, from his accustomed goodness, has been also pleased to give up to the Society the upper garden belonging to the Chateau, for the purpose of making experiments, and of raising fruit trees; it is put under the management of four of the directors conversant with gardening, and agriculture, whose zeal it is not doubted, will be attended with public advantage."

"The prejudice which the agriculture of the province has hitherto sustained, from the custom of abandon, from the expense attending the recovery of damages for the trespass of cattle, of the breaking down, carrying away and destroying fences, and of settling disputes between neighbours, respecting boundary fences, and ditches, in all which cases, as the law now stands, it is necessary to bring actions in the Courts of

Common Pleas, when in general the damages recovered will not compensate for the expense and attendance in the Courts; all those matters did not escape our notice, to remedy them, was thought worthy of the attention of the Directors, and a committee was named to draw the project of a law for the purpose, which was submitted to the Legislative Council; their ideas on the subject, however, differed from those of the directors."[63]

It seems that the president was too pessimistic regarding the matter of an act, for one was passed at that time abolishing the custom of abandon and requiring the enclosing of stock at all times and providing penalties for destruction of fences, crops, etc.

The Society also supported the government efforts to increase the production of hemp. Weld outlined the advertising procedures that were used in an attempt to stimulate production but stated:"It is not probable that much will be raised for a considerable time to come."[64] In this he was quite correct for, in spite of the various inducements that were offered from time to time, hemp production never did gain favour with the farmers.

The activities of the Society apparently had a secondary effect for a number of letters and articles about various aspects of agriculture appeared in the papers during the next few years. Most of these concerned crop production but the Society itself also paid some attention to the improvement of livestock for in June 1791 a notice appeared that: "The Agricultural Society has purchased a beautiful bull of the Blackwell breed, which is esteemed one of the best in England both for beef and milk. He is kept at Mr. Lynd's farm for the use of every one who wishes to improve the breed in this country."[65]

Direct encouragement of increased production was another endeavour of the Society and in April 1796 the following notice appeared: "At the quarterly meeting of the directors Sept. 18, 1795, Resolved that a premium of thirty dollars be paid by the treasurer to the person, being a Habitant, who shall raise the largest quantity of merchantable wheat from a superficial arpent of land and not less."[66] The notice went on to specify $25.00 for the next largest amount and $15.00 for the third. For potatoes $12.00 was offered for the greatest quantity grown on one half arpent, with $8.00 for the second and $4.00 for third. For winter wheat, in 1797, the prizes were $40.00 and $20.00, and for flax $12.00, $8.00, and $4.00.

The overall success of the societies is hard to measure but Jones gives the following evaluation. "Still another example might be found in the unhappy history of the Lower Canadian agricultural societies. The British farmers who promoted these societies and their exhibitions tried to interest the French-Canadians. They met with scant success, for in the societies where the French and British farmers competed on a basis of equality, the British won all the prizes, because they had superior farming knowledge, and the habitants soon withdrew in discouragement; in the other societies, which had separate classes for the British and French exhibitors, the directors found they were giving prizes to the habitants for rubbish. Finally, the habitants failed to benefit even from observing the practices of the British who were settled in their midst. Under these circumstances the hopelessness of any program of government agricultural education which might have been undertaken is understandable."[67]

This rather gloomy picture of the general state of agriculture, and it must be remembered that the habitants outnumbered the British at least ten to one, was supported by other observers. Thus Campbell, in 1791-92, noted: "The quantity of wheat annually shipped for Britain is very considerable, and yet no sort of manure is ever used, as formerly observed. The dung is laid upon the ice so as to be carried off by the floods of spring. The Canadians are perhaps the worst farmers in the world. If one of them happens to have a spot in a field that produces nothing, and had industry enough to drop a few carts of dung on it, if the plough and harrow do not spread it, it may lie there for him; he has done his part when he took the trouble of putting it on the land; spreading it is a labour no one would submit to undertake. Their horned cattle are perhaps the worst in the world, and their horses the best; the former are ill made, big bellied, thin quartered, and poor as carrion, though this season they ought to have looked best. While the latter are plump, round, well made, stout and full of spirit. I have seen Horses of many nations, but none in my opinion for common service equal to those of the Cameraskas in Lower Canada."[68]

About 15 years later Lambert[69] could see no real improvement and pointed to the same basic weaknesses noted by Campbell. The only item of praise he had was for the poultry which he considered to be quite good.

An encouraging note was sounded by Abbé Bedard, cited by Ouellet. His observation was that where the French Canadians lived in homogeneous settlements they continued with their traditional ways, whereas when in mixed settlements with the British there was a tendency for them to change their practices and emulate their British neighbours. In this Bedard differed from the observations of Jones noted earlier.[70]

Sansom put his finger on one of the underlying causes of the relative backwardness of the habitants when he noted: "Most of those who cultivate the soil can neither read nor write, of course they know nothing of the advantages of composts, or the rotation of crops, by which the means of life are so cheaply multiplied by the intelligent agriculturist."[71]

Palmer, visiting in 1817, saw a slightly better picture than some of the earlier observers, but may have been viewing an area with a greater English population. He wrote: "The country, after leaving the banks of the Sorrel, is level, but not swampy, as far as the eye can reach. It is populous, and divided in right lines by ditches or rail fences, into many

Quebec

luxuriant farms. The land is excellent, much of it in meadow, and more potatoes than I have noticed growing in the same distance anywhere in America. Indian corn comes to little beyond 44 or 45 N. lat."[72]

Bouchette[73] also gave a much more favourable picture, but he was reporting on new settlements which were occupied primarily by English-speaking settlers.

One interesting development, late in the 18th century, was the appearance of the first contract production of crops. In December 1791 there appeared an advertisement, by the Quebec Brewing and Distilling Company, for contract production of barley. The company was to provide seed at prime cost and guaranteed a price of two shillings and sixpence per minot of fifty-four pounds, English weight, at their works or at any other place, less shipping charges.[74]

Newspapers provide an insight into the level of knowledge that was available to the farmers of the time. One item dealt with a problem that still is of concern to livestock producers, namely bloat in cattle. "The death of cattle which often happens in consequence of eating too freely of fresh clover, may be prevented by giving them, by means of a bottle or horn, a large handful of common salt dissolved in a pint of water, or which is perhaps better, mix three gills of sweet oil with one pound of butter, to be given as above directed. If this does not produce the desired effect in less than half an hour repeat the same quantity, walk the cow gently about during the time."

For fodder conservation the following advice was offered: "To preserve clover hay and improve the quality of straw, in a day or two after your hay is cut, when only about half dried let it be taken in and packed with alternate layers of straw; giving each layer of clover four or five half pints of salt, or more, in proportion to the quantity of hay; three pints to the hundred weight will prevent fermentation or excessive heat, which injures and moulds it. By not drying the hay as much as is common, and putting it up as above mentioned, it imparts to the straw the flavour of the clover, and cattle eat one as greedily as the other."[75]

Advice on growing corn was offered as follows: "It is a fact, which those who cultivate Indian Corn would do well to attend to, that the oftener the ground is stirred in Cornfields, the more secure the Corn is against drought — as frequently changing the surface of the soil prevents it from becoming so dry and hard as it otherwise would; and keeping the weeds well under, preserves to the Corn much nourishment which the weeds would draw from it if suffered to grow. If this fact is doubted, observation will remove all doubts on the subject."[76]

The above elicited a fairly long dissertation from "Agricola" (refer to Chapter 2, page 41 for details on Agricola) in Montreal on cultural practices for corn. The final paragraph is of particular interest: "An acre is believed to produce from forty to fifty bushels, if well attended, but from the little care that is taken, to estimate the return of crops is generally guesswork. However, I have to recommend hoe crops, in order to destroy weeds, so prevalent in this country, which if not attended to, will, ere long, ruin many families, and I hope ingenious gentlemen will take steps followed by the first personages in Great Britain, to encourage agriculture."[77]

In 1807, the *Quebec Gazette*, published a pamphlet from Philadelphia, giving the history of the Hessian fly that had been so destructive to wheat, outlining two seed treatments for the control of the pest. One method was heat treatment in the oven to destroy the larva, but not the germinating ability of the wheat. The other was to steep the seed in a lye-quicklime solution.

Advice also was offered on financing: "Do not run into debt to buy land. Land will not generally support a family, and pay taxes and interest on its value. If you have but a small piece of land, cultivate it well, make it produce as much as possible; and if you can get more than will maintain you from this little farm, lay out the surplus in buying more. If you cannot get more than a subsistence, it is time to think of lessening expenses, or selling out and buying new land. Depend on it, farmers who pay interest do not work right."[78] Whether that was good advice at the time is not known, but it is hardly in keeping with agricultural practice today when heavy capital investments are required.

Finally, on the subject of fertilizers, there were a couple of suggestions. First: "I would beg leave, through the medium of your paper, to solicit my brother Farmers to make use of their Ashes on their Corn Land, instead of selling them to the merchant for 12 cents a bushel. I will warrant them 35 cents a bushel, clear of all expenses, if rightly managed."[79] The writer then went into considerable detail on how best to use ashes. There also was an item on the value of dung as manure, including the suggestion that the use of green dung was responsible for many of the insect problems and to overcome this the dung should be stored for a certain length of time. The use of Plaister of Paris as fertilizer also was given attention.

Apparently worms were a problem on fruit trees and two methods were proposed for their control. The first was: "Take a pound of good potash, and dissolve it in one gallon of rain water, and then put half a pound of quick lime in it and boil it half an hour, take it off and set it to cool and it will be fit for use. Make a mop of linen rags, with a handle long enough to reach the lower limbs of the tree, then wash the trunk of the tree with the lie, from the lower limbs to the ground; and the sharper and more pungent the lie is, the better."[80]

The second treatment was given the force of law in "an act for the preservation of Apple Trees in the Parish of Montreal." The insect was identified as the caterpillar Arpenteuse and the act stated: "it has been found by experience, that the ravages of that insect can be materially checked, and the number thereof diminished by tying in the spring round the stem of each Tree, a bandage, and smearing the same repeatedly with Tar, so as to present an obstacle to the progress of such insects from the ground to the branches, by which means many of the Trees that otherwise would have been destroyed have been preserved..."[81]

The worms on fruit trees were a relatively minor problem compared to other environmental hazards to which the crops were exposed from time to time. In the early French period it was noted that various hazards were encountered, namely, frosts, grasshoppers, caterpillars, etc. Later on, occasional references were made to specific problems such as cutworms that cut the roots of young tobacco plants and forced successive replantings to the point where the crop would not mature. Crop failures were recorded in a number of years, no doubt caused by some of these factors.

As agriculture expanded the number of problems seemed to increase. Mention was made earlier of the concern with smut in wheat. This crop was the most important crop, both from the standpoint of nourishing the local population but, increasingly, as an important export crop. Wheat production expanded greatly in the early 19th century and exports rose from 3,106 bushels in 1796 to 1,101,033 bushels in 1806. But trouble for the wheat producer was developing. Ontario and the United States were able to supply wheat of superior quality to that produced in Quebec and, despite greater transportation cost to overseas markets, were strongly competitive. This competition, and especially periods of low prices caused by surpluses from these sources, was partly responsible for the decline of wheat production in Quebec. But the major factors in the decline were reduced productivity caused by disease, insects, and soil exhaustion.

There is little doubt that a significant factor was the appearance of the Hession fly. This pest made its first recorded appearance in 1805 and was prevalent and destructive for a number of years. But worse was yet to come from the ravages of the wheat midge. This insect was noticed first in western Vermont, in 1820, and may have been in Lower Canada at that same time, but first reports of serious damage came in 1828, 1829, and 1832. In 1834 it first appeared in numbers in the Montreal area and in succeeding years caused extensive damage throughout the area. Added to the insect problem was the wheat rust, which in certain seasons was equally destructive.[82]

The almost complete destruction of the wheat crop year after year created a real problem for the farmers. "In an endeavour to find a substitute for the wheat flour which they would have to purchase, they began to depend in large part on potatoes for food. Then, in 1843, the potato crop in Lower Canada, as throughout North America in general,

The Trail Blazers of Canadian Agriculture

was attacked by the 'late blight' or 'rot'. The disease was exceptionally bad in 1844, 1845, and 1846, and by the end of 1847, the crop had been given up in large sections of Lower Canada."[83]

Faced with a depressed agricultural industry, the Legislative Assembly set up a special committee to study the state of the industry and make recommendations for correcting its ills.[84] The committee reported that agriculture had made great progress during late years, though how it arrived at this conclusion is a little difficult to understand in light of the conditions that had caused the committee to be set up. It admitted that "the soil went on getting poorer until having lost all its strength it ceased to produce wheat, or produced only a sickly grain without the strength to resist accidents."

The members suggested that one of the main problems was the lack of agricultural education of the farmers. Other defects were: "want of improved instruments, the insufficiency of drainage in certain Districts, the complete destruction of our forests, part of which should be preserved for shelter and part for sugaries. The want of attention shown by the Legislature on this subject; the want of agricultural education, and the absence of a sufficient market, was also pointed out."

The recommendations of the committee are of considerable interest as they marked the first official recognition of the need for agricultural education and extension work. The main recommendations were: "Agricultural Societies, such as already exist; Model Farms, with Schools of Agriculture; the publication of Elementary Treatises to be distributed gratuitously among the population of the country parts, and the schools; the publication of a Journal, and the appointment of two Superintendents." The latter were to act as liaison officers and also as lecturers to public meetings.

With regard to Schools of Agriculture the committee recommended specifically: "Your Committee, therefore, suggest that a special annual grant be allowed to each of the Colleges of St. Hyacinthe, L'Assomption, Nicolet and St. Anne, on the condition that a Chair of Agriculture be established for the instruction of their scholars, and that a portion of land, in the immediate vicinity of each institution, be cultivated as a Model Farm. A similar grant might be made for the same purpose in the Townships, at one of the Academies where a portion of the youth who speak the English tongue receive their education."

The committee then outlined the current expenditure on agriculture and recommended a change in the distribution of funds to provide for implementation of their recommendations without any increase in expenditure.

Current expenses:

For 36 counties, at £150 each	£ 5,400
Three Districts, entitled to an annual grant of £500	1,500
Annuity to the Lower Canada Agricultural Society	600
	7,500

Proposed expenditures:

For prizes to be granted by the public Exhibition Societies a sum of, being about £100 for every 20,000 souls	4,000
For five Schools of Agriculture, with Model Farms attached to Colleges and Academies, to be distributed in equal portions	1,500
Premium to the author of the best elementary treatise in both languages	600
Annuity to the Lower Canada Agricultural Society for continuing the publication of a journal	700
Salary for two Superintendents, including their travelling expenses	700
	7,500

The committee made one further interesting recommendation subject to additional funds being made available. This was: "and grant, in different parts of Lower Canada, an annual sum of £200 to some good farmer, possessing a good farm and a sufficient number of cattle, together with the advantages of an elementary education, on the condition of his cultivating his farm as a Model one, under the direction of the Superintendent of his District, and obliging him to show and explain to every visitor the details of his model cultivation. This sum of £200 added to the means already in his possession, would enable him to improve his system of cultivation and his breed of animals, and to procure instruments of a superior make, at the same time, that it would allow him to dispose of a portion of his time in explaining the details of his art to his visitors."

This concept of demonstration farms had been pioneered by Talon, many years earlier, when he reserved lots, in new villages, for experienced settlers who could give advice to the newcomers. However, the concept had not been developed to the form now envisioned. The concept has been applied widely since that time, throughout Canada, with a great range of variation in detail.

The Lower Canada Agricultural Society prepared a special report to the Legislative committee.[85] At the same time it circulated a questionnaire to its members with a series of questions relative to problems of production. The following sample of questions portrays quite clearly the type of problems that were of concern.

"What is the cause that a greater number of sheep are not raised in Lower Canada?

What is the most profitable mode of raising horses for market? What would be the most suitable and profitable method of raising meat cattle in Lower Canada?

Can these animals be supported during the winter advantageously on straw alone?

Is it advantageous to keep milch cows?

Which are the most profitable breeds of swine for Lower Canada, and what is the most profitable mode of raising and fattening them?

Which would be the most suitable variety of wheat to sow in Lower Canada?

What is your opinion of fall wheat?

Would you recommend the extensive culture of the potato?

What are the weeds prevalent in your part of the country and what means are taken to destroy them?

As one might expect, there was a wide range of answers to each question. With reference to livestock the replies indicated that the Leicester was the most popular breed of sheep. Durham cattle were most popular for crossing for beef, though the Hereford was mentioned. Devon and Ayrshire cattle were the most popular for milk. The Canadian horse was in high favour. The Berkshire breed of pigs was most popular, but Yorkshire, Chinese, and Ohio breeds were mentioned.

There was no strong support for potato production because of the disease called rot. Black Sea wheat was the preferred type, in preference to the old 'blé blanc' variety. Fall wheat was considered to be too uncertain a crop.

The list of weeds is interesting as it includes many that still create problems: Canada thistle, pigeon weed, barweed, chicken weed, barn grass, wild peas, mustard, Boston grass, buttercup, large and small daisies, twitch or catch grass, yellow rocket, and wild oats.

Considering the low level of use of manure it is of interest that one respondent suggested that when a farmer could not get sufficient manure he should plow in green crops such as clover, buckwheat, or peas.

Legislation Enacted

In Quebec, direct government intervention in agriculture began in 1659, when the government of France took direct charge of the colony. At that time several measures were introduced to encourage agriculture. Among these were an increase in immigration and the introduction of a new supply of livestock. In addition, various ordinances were passed to regulate production and marketing. After the British conquest further action was taken to regulate and support the industry. Examples of official actions in the following list of proclamations include acts of both the French and the British.[86]

1665
- An ordinance prohibiting any person from pasturing animals on land they did not own.

Quebec

1667
- An act fixing the toll a miller might take for grinding wheat.

1676
- An act making it an offense to trespass on seeded land.

1701
- A decree fixing the price of grain, and ordering the inhabitants to offer it for sale. The decree applied to wheat, Indian corn, and peas.

1707
- An act making it an offence for anyone to enter on land, not his own, and take fruit.

1708
- An ordinance ordering the country people to bring their commodities to the market place of the Lower Town, on market days, and not expose them for sale on the shore.

1709
- An ordinance making it an offence for the inhabitants of Quebec to let their vicious dogs roam at large as these dogs devoured the settlers' sheep.
- An ordinance prohibiting anyone from hunting on seeded land.

1710
- An ordinance forbidding the breaking down of fences, or destroying trees.
- An ordinance ordering owners of swine to put collars on them.
- An ordinance prohibiting swine from running at large on the street.

1727
- An ordinance forbidding millers, farmers, and others from selling their flour, peas, or grain otherwise than in sacks.
- An ordinance regarding the practice of abandon of cattle in the autumn.

1731
- An ordinance ordering inhabitants to have their swine ringed during the summer.

1733
- An ordinance prohibiting the removing of any wheat or flour from within the colony of Montreal, before the end of seeding time, and the removing thereof, after that time, without permission.

1738
- An ordinance fixing the price of wheat at four livres per minot.

1741
- A new ordinance respecting the abuse of allowing swine to run at large throughout the town of Montreal. Permission was given to anyone to kill such swine and appropriate the meat.

1742
- An ordinance prohibiting the selling of wheat flour at higher than the fixed price.

1761
- An ordinance respecting the enclosure of hogs. Permission was given to anyone to kill swine running at large. This indicates that the British encountered the same problems that the French had.

1764
- An ordinance to prevent the forestalling of the market, and frauds by butchers, etc. This ordered that all produce (except oxen and sheep) brought to Quebec, Montreal and Three Rivers had to be offered for sale at the public market.

1765
- A proclamation making available grants of land at two shillings annually, per 100 acres, after the first two years. For every 50 acres granted, three acres were to be cleared and worked within three years. The grants were for 100 acres for any head of family plus 50 acres for every other member of the family.

1770
- A proclamation banning the export of cattle and other livestock and flour.
- A proclamation requiring owners of grain to have it threshed and ready for transportation to a place of security. This was due to possible threat of invasion from the U.S.
- Proclamation placing an embargo on the exportation of corn, grain, flour, biscuit and salted provisions.

1793
- An act establishing the Winchester measure and a standard for other weights and measures throughout the province.

1802
- An act for the encouragement of the culture of hemp in this province. A sum of £1200 was appropriated for this.

1804
- An act regulating the curing, packing, and inspection of beef and pork to be exported from the province.

1805
- An act for the relief of the poor in the loan of seed wheat, corn, and other necessary grain. This followed a crop failure the previous year. Similar acts were passed in 1811, 1817, and 1829 after crop failures.

1805
- An act for the preservation of apple trees in the parish of Montreal.

1818
- An act for the encouragement of agriculture in the province. This provided up to £2,000 for agricultural societies with £800 to Quebec, £400 to Three Rivers, and £800 to Montreal. This act was extended and modified from time to time, but continued to give support to the societies.

1819
- An act to secure the inhabitants of the Interior District of Gaspé in the possession and enjoyment of their lands.

1824
- An act for the more speedy remedy of divers abuses, prejudicial to agricultural improvement and industry in the province, and for other purposes. This was an omnibus bill that prohibited the breaking down of fences, the running at large of stallions and rams, provided for fence viewers and regulated fences and drains, provided for animal pounds, and provided for the destruction of noxious weeds.

1829
- An act to appropriate a certain sum of money therein mentioned for the encouragement of agriculture. This was the first year when reference was made to the use of some of these funds for prizes for exhibitions.

1831
- An act to encourage the destruction of wolves.
- An act for preserving, for the purpose of husbandry, the grass growing on the beaches, in the district of Quebec.

Despite the problems besetting agriculture in Quebec during the first half of the 19th century, the industry had developed and was an important element in the economy of the province. This is evident in the data in Table II which also provides comparative data for 1966 to show the growth of the industry in the intervening years. It is evident that there have been major shifts in the relative importance of various items of production. The dairy industry, which was just beginning to be an important factor in the mid-1800s, has become a major industry in the province. The decline of wheat production, noted earlier, continued to the point where this crop is no longer important whereas oat production has increased to provide feed for the increased livestock population.

Tabel II
Periodic population and agricultural statistics for New France, Lower Canada and Quebec during early development, with 1966 data for comparison[a]

		1667	1695	1734	1784	1831	1851	1966
Population	No.	3,918	12,786	37,716	113,012	553,134	890,261	789,000
Cultivated land	Acres	11,448	28,110	163,111	1,569,818	2,066,213	3,605,167	7,629,346
Horses	No.	--	580	5,056	30,146	116,686	148,620	62,138
Cattle	"	3,107	9,181	33,179	98,591	388,706	591,552	1,797,646
Sheep	"	85	918	19,815	84,696	543,343	648,665	112,438
Swine	"	--	5,333	23,646	70,465	295,137	256,794	1,173,687
Wheat	Bus.	--	129,154*	737,892	--	3,407,756	3,073,943	298,000
Barley	"	--	20,710*	3,462	--	1,074,866*	495,768	526,000
Rye	"	--	--	--	--	--	325,422	69,000
Oats	"	--	13,955	163,988	--	3,202,274	8,977,400	41,940,000
Buckwheat	"	--	--	--	--	--	532,412	16,000
Peas/Beans	"	--	--	63,549	--	984,758	1,415,136	66,000
Corn	"	--	6,490	5,223	--	--	401,284	--
Potatoes	"	--	--	--	--	7,357,416	4,429,016	7,239,000
Hay	Tons	--	--	--	--	--	755,579	4,132,000
Butter	Lbs.	--	--	--	--	--	9,610,036	133,860,000

* Includes other grains

a. Data for the years up to and including 1851 are from *Census of Canada* 1665-1861. Taylor, I.B. Ottawa 1876 Vol.4. Data under the heading 1966 are from the 1966 census for the first six items. The remainder are 1965 data from the 1967 *Canada Year Book.*

Chapter VI
Ontario

European agriculture was introduced into Nova Scotia, New Brunswick, Prince Edward Island and Quebec by the French. In each of these provinces agriculture had become relatively well established before the English took possession. Ontario reflected a different situation. Here there were essentially three thrusts, two by the French and one by the English, and none of them led to any major development prior to the end of the French regime. The only one of the developments that might be considered to be continuous and of any significance for the future was along the shores of Lake Ontario and Lake Erie, where small French settlements were in existence when the English took possession of the country.

It is difficult to pinpoint the exact beginning of European agriculture in Ontario. The Huron and Tobacco Indians, in the area east of Lake Huron, had a fairly substantial agricultural industry when the Europeans first arrived. Their crops were mainly corn, beans, and squash. One estimate is that, at one time, they had as much as 23,000 acres under cultivation. Though this seems to be a high estimate there is no question that their agricultural activities were extensive.

It was into this area and this situation that the Recollet and Jesuit missionaries made their entry in the second decade of the 17th century and where they began to exert an influence on agriculture. While it is known that the first Recollet missionary, Le Caron, went to the Huron Indian country in 1615, the exact date at which European crops and domestic animals were introduced has not been determined. It is known that by 1637 poultry had been introduced as we find messages from the missionaries verifying this. Thus: "We had, indeed, one hen, but she did not every day give us an egg."[1] Also, Father Mercier, in conversation with an Indian, had the following exchange: "My friend, I pray thee, speak a little lower." "Thou hast no sense," he said to me: "There is a bird, speaking of our cock, that talks louder than I do, and thou sayest nothing to him."[2]

The first reference to wheat being grown appears in a report for 1638: "We gathered our little harvest and our vintage for the holy Altar, in the month of September. The harvest was about a half bushel of good wheat, which was large for the little we had sowed, and a small keg of wine, which kept very well during the entire winter, and is still passably good."[3] The following year reference again was made to grain in the report of Father Peron: "We have no land of our own, except a little

The Trail Blazers of Canadian Agriculture

borrowed field, where French grain is raised just to make the host for the holy mass..."[4]

The first reference to domestic animals came in 1646 with the following: "Caron, who was taking some calves to the Hurons, left Three Rivers on the 11th of May..."[5] Another reference was made to this in September of that year: "This last man (Eustache Lambert) had given himself in service and was to go up again, and he did indeed go up again with the above named persons, and, moreover, he took charge of two calves."[6]

Just how the animals were handled on the trip to the Huron country has not been reported. Considering the distance from Quebec to the Huron country, the fact that the canoe was the means of transportation, and the difficulties of travel, it could not have been an easy task. "The good people now see in their country another kind of animal, of which they have never had any knowledge; there are little bulls and heifers, which have been brought hither with great labors."[7]

Pigs also were brought in at some time, for in 1649 Father Ragueneau reported: "For we have larger supplies from fishing and hunting than formerly; and we have not merely fish and eggs, but also pork, milk products, and even cattle, from which we hope for a great addition to our store."[8]

Considerable progress in developing an agricultural base at the missions had been made, but then disaster struck as the Iroquois attacked and destroyed all of the Huron settlements and drove the inhabitants out of the area. Some of them settled on an island in the lake and, in 1650, the missionaries reported: "We saved ten fowls, a pair of swine, two bulls, and the same number of cows — enough doubtless to preserve their kind. We have one year's supply of Indian corn; the rest has been used for Christian charity."[9]

Some appreciation of the scope of the development that had taken place prior to the devastation of the area is given by Eccles:

"In 1634 the Jesuits established St. Joseph mission in the village of Ihonatoria and three years later a second mission at nearby Ossossane. Then, in 1639, a much more ambitious establishment, named Ste. Marie, was set up near the mouth of the Wye River. It consisted of a chapel, hospital, mill, stables, barns, a residence for priests and another for lay workers, all surrounded by a log palisade with stone bastions. As many as sixty-six French resided there at one time, priests, lay brothers, servants, a surgeon, an apothecary, and a number of artisans. Cattle, pigs, and poultry were brought by canoe from Quebec, and the fields were cleared to grow wheat, corn, and vegetables, sufficient to feed the hundreds of Indian converts who eventually came to settle nearby."[10]

The destruction by the Iroquois in 1650 was complete and ended any real agricultural development in that part of the country until after the arrival of English settlers over a century later.

Some 20 years after the above-noted events there was a new agricultural intrusion in an area most unlikely for success. This was in the James Bay area where the Hudson's Bay Company was establishing trading posts. The Governor and the Committee of the Company seemed to be well aware of the need for an adequate, economical food supply for the employees stationed at the various posts and thought to assure this by agricultural development at the posts. While no commercially viable agricultural industry has developed in the area, the extent of the efforts made, both in time and scope of production, are of considerable interest and will be dealt with in some detail, indicating some of the misconceptions held by the people in London, their stubbornness in continuing the efforts in the face of the poor results, and the difficulties encountered by the employees in the field.

The first reference to agricultural activity is a Minute of the Committee on May 14, 1674: "That 6 bushells of bay Salt & 2 bushells of white Salt, in all 8 bushells bee Sent besides what hath been formerly ordered, & 4 bushells of the Several Sortes of graine following bee alsoe provided for the Countrey, viz. of wheate, rye, barley & oates."[11] This was followed by a Minute on May 16: "Ordered that there be provided a bushell of wheate, & of rye, barley & oates, or a barrell of each in caske, Such Sortes of garden Seedes as the governours Shall advise."[12]

In 1680, instructions were sent to Governor Nixon at Moose River as follows: "Wee have now sent you severall sorts of seeds which you will find in the Invoice as wee have formerly done to enable you to make the Experiments, and we doubt not but Fish & Fowl may in the proper seasons be catched in considerable quantities, and kept salted & dryed for the relief of the men. And upon Hayes Island where our grand Factory is, you may propagate Swine without much difficulty, which is an excellent flesh, and the Creature is hardy and will live where some other Creatures cannot. And upon so little a spot of ground as that island is you may with your men who will be constantly attending the Factory, and hunting up & down, guard them against Bears, Wolves or any other Creatures of prey that may come over upon the Ice from the Mainland. So that wee think you might be in a great measure supplied with that sort of provisions with reasonable care, and wee therefore desire you to be diligent therein."[13]

To make sure that action would be taken they sent stock: "Wee have sent 1 he Goate & 2 she Goates & 1 sow with pigg which we have done in hope they will increase in the Country & be of use & comfort to our people which is a thing that deserves your utmost care as well for the Good of the Factory as for the ease of the Compa. in the businesse of Provissions."[14]

In the list of shipments outward at that time we find:

3 bush hempe seeds,	2 papers turnip seeds
2 peck hempe seeds,	one bushel hempe seeds

The Trail Blazers of Canadian Agriculture

and turnip seeds.[15]

Reporting back to the Committee in 1682 Nixon had both good and bad news: "...so that these things being over, we have nothing els to doe but to provide for our winters provisions of what the country can afford us, it being then the tyme of the yeare which if made use of, may be a great help to the saving of our English victuals and with what our gardine produceth, for at that time we may probably save 2 hogs-heads of gees, and 3 or 4 hogs-heads of turnips, besides colewarts, all of which are refreshing to the men..."[16]

More good news was: "Things so observed, and our business to be most at home we can make hay in the tyme of yeare, to feed a good heard of goats, and look after our gardines, and try whether corn will grow or not, I am in good hopes that Buc-wheat, will growe by reason of the quickness of growth thereof wherefor I desire that you send me a little for a tryle."[17]

The bad news was: "One thing more I must advertise you that your Honours goats are dead, and that it is lost labour to trouble yourselves to send either them or hogs for they destroy more provisions than they are worth, but if anything thrive it will be shotland or Orkney sheep for they have good coats on their backs to defend them in the winter from the could, in the summer against flyes."[18]

There were not many reports of real success but there are indications that some encouraging results were reported. For example, in 1694, the Committee wrote to the governor at Albany: "Wee are glad to heare that the Bottome of the Bay is a fertile & rich Country as you express it, however wee have not sent any Cattell this Yeare by reason you Intimate the great difficulty & danger in lookeing after them dureing the Warr."[19]

In 1696 they were still attempting to get wheat and hemp grown: "We have sent you the seeds you desire as you will find by the Invoice as allsoe seed Wheate & though you mention that neither that nor hemp seed will thrive with you (but flax & barley, very good) wee apprehend that after 2 or 3 crops of Barley, if you sowe hemp or wheate upon the same land, it may be much better then you expect the cause of the hemp growing soe gouty arising from the richness of the Soyle. Wee desire that you goe on to Save Flax send us what you raise & keep a stock of seed."[20]

Some idea of the extent of the livestock operations is given in correspondence from Fullertine at Albany in 1703: "I understand you are glad to hear of the increase of your flock of Cattle which is now swelled to one hundred and eight. I shall do the utmost of my power to preserve the stock by keeping the females... is without prejudice to the Factory, for truly I must needs say that they are no ways fitting to be keep'd in the Country, in time of ware for in the most Dangerous time in all the year, that is from the first of July to the 10th of August half my hands were fain to ly out to gett them hay, to ye Great hazard of ye whole Factory & the year the Ferry went last home allthough I had two cocks of hay left & 12 or 15 men continually out, could hardly get enough for them & this year my stock is much larger. I desire to kill most of the hees this year which will go butt a very little way to Serve out amongst the men for they are so prodigiously small, especially the Goates. Yet there is many of them of a year old yet I'm confident that there are severall men in the Factory could eat up one of them in a day, the Sheep are not to be complain'd of but thrive indifferently well."[21]

Three years later another governor at Albany wrote: "I have kept in the Country your flock of Cattle consists of 102 but I design to keep but forty which shall be the best breaders."[22] It is not clear what class of stock was kept in that the term cattle apparently was a generic term that applied to all classes except horses and swine. It is clear that sheep and goats were involved, but there is no specific reference to cows. The only suggestion that cows were kept comes in a diary entry for May 15, 1706: "Our people has been imployed to day a carrying the Dung out of the Cow house."[23]

One problem that they encountered is indicated an entry for May 29: "one of my Old Ewes that was brought from England and another which was bred here in the Countrey Died to day, which is usuall when they come first to feed upon Grass after so long being kept upon Hay that the alteration of foode purges and Scurers them to the degree that they become very weake. We Loose more or Less of them every Spring."[24]

At a much later date livestock still were being maintained at the posts on James Bay, for a diary kept at Moose Fort had the following entry for November 21, 1783: " And afterwards went up the River to fetch home the Cattle, they returned with only two Cows, not being able to get the others home."[25] The next day they were more successful: " The men brought home the remainder of the Cattle."[26] Other entries indicated that the cattle were used for hauling firewood as well as to supply beef. Some concept of the size of the herd can be gained from the entries of the following year: "Nov.6. — Gill and one man killing 4 hogs....and fetching home 16 of the Cattle all they found. Nov.9. — A. King brought home 4 more of the Cattle. Nov.10. — Edward Clouston the cowkeeper, and two other Men got home the remainder of the cattle."[27]

Much more could be recorded about the problems encountered in the continuing efforts at agricultural production at the posts but enough has been given to provide a picture. Despite the persistent efforts, the testing of various possible crops, and the attempts at the production of several classes of livestock it is clear that neither the climate nor the soil conditions were conducive to any major development in this area. With changes in the fur trade during the first half of the 19th century, the posts on James Bay lost their importance and agricultural activity ceased.

One amusing incident should be related before this part of the Ontario story is concluded as it shows the limitations that prevailed and, at the same time, that adaptations were made. The story was related by John Tod who spent some time at Fort Severn: "In the fall of the year, a horse was sent out from England - from the Orkney Islds, to York Factory, to one of the Cos. servants. On the arrival of the ship, it was too late to send the animal to the interior. Here the poor beast was considered lost, as there was no hay for him, nor anything else, for that matter. At York they took it into their heads to send him on to Mr. Tod at Fort Severn, some 250 miles along the coast."

"I was indignant, as I had no more means of keeping the animal. A week after the snow set in, & the weather became terribly rigorous. There was no grass to be had; no moss either even had the snow not covered it over; & the only companion had the poor horse, was a great big boar. How he came to be there I cannot tell. But the hog was taught by experience to subsist on fish, and preferred them (there was no choice) frozen. The horse at last took it into his head to try fish too."

"The fort had a bell to ring the men to work, to their meals etc. So used did the 2 animals become to this signal that they were generally found to be sitting quietly before the door waiting for some food, on the occasion of meals. The refuse of the meals was put into a pan and given the horse and his companion the hog. They would rush for it, and at the last morsel, there was always sure to be a fight over it. By this means the horse was kept alive the 7 months of winter..."

"I had to proceed every summer to York on business; rode the identical horse thither & every one was astonished to see him live through the terribly severe seasons in that latitude."[28]

Continuous Production in Southern Ontario

Attention now is turned to southern Ontario where some continuity of production has occurred since the first attempts at agriculture in that area. The first action was at Fort Frontenac, at what is now Kingston. This post was established in 1671 and in the early years of its existence some agriculture developed. Father Hennepin, who served there as missionary from 1676 through 1678, gave the following account:

"The Ground lies along the Brink of this lake is very fertile. In the space of two years and a half that I resided there in discharge of my Mission, they cultivated more than a hundred acres of it. Both the Indian and European Corn, Pulse, Pot-herbs, Gourds, and Water-melons, throve very well. It is true indeed, that at first the Corn was much spoiled by Grasshoppers, but this is a thing that happens in all Parts of Canada at the first cultivation of the Ground, by reason of the extream Humidity of all that Country. The first Planters we sent thither, bred up Poultry there, and transported with them Horned Beasts, which multiplied extreamly. They have stately Trees, fit for building Houses or Ships. Their Winter is by three Months shorter than at Canada. In fine, we have all reason to hope, that ere long, a considerable Colony will be settled at that Place. When I undertook my great Voyage, I left there about Fifteen or Sixteen Families together...."[29]

Jones puts the development in perspective when he notes: "It should be added that in another work Hennepin noted that La Salle had only thirty-five head of cattle at Fort Frontenac. This agricultural enterprise ended in a few years when Denonville abandoned the fort. When it was rebuilt later, there was again some farming. Charlevoix, who visited in 1721, found that the island opposite the fort was known as Hog Island, because so many pigs were kept there. But Fort Frontenac remained a military and trading post and little more. When Rogers went up the Great Lakes in the autumn of 1760, he noticed that five hundred acres or so of cleared land around it were overrun with clover and pines."[30]

The next westward thrust was to Detroit where a post was established in 1701. While this was in what is now the United States it soon led to development on the Canadian side of the river. Though this post, like all of the others, was established as a trading and staging post, it also was to become an important supply centre for more western posts. To this end the government of New France took steps to encourage settlement and the development of agriculture. Glazebrook outlines the inducements that were made to prospective settlers in the older settlements of Quebec.

"The terms were generous enough; the land itself, food for a year and a half, agricultural implements, domestic animals, roofing nails, powder, and lead. Those who accepted the offer were a mixed lot. Some came from around Montreal and these included voyageurs who already knew the area. There were also applicants from Detroit itself, disbanded soldiers, Africans or panis (Indians)... It is estimated that at the time of the American Revolution a hundred families lived on the Ontario shore... A feature of the area was the number of orchards growing pears, peaches, plums, and apples."[31]

Reaman expands this description: "In that area, the farmers settled in French fashion along the bank of the river between Sandwich and Amherstburg, and the community was called Petite Cote which had some fifty families about 1750. Each family had about two fields in which they alternated crops, mainly spring wheat and peas. They grew wheat, barley, oats, peas, buckwheat, Indian corn and potatoes."[32]

Even while these developments were taking place the next step toward the west was being taken into an area more favourable than at James Bay but not so favourable as at Detroit. La Verendrye, fur trader and explorer, established a post on the Lake of the Woods in 1732 and in 1733 reported: "There is good fishing and hunting, quantities of wild oats (wild rice), and excellent land cleared by fire which I am now putting to seed."[33] In referring to an Indian chief he wrote: "He told me that he would remain near the fort all summer with the elders of his people to defend us, and that he was going to raise corn as we do. I urged him to raise as much as possible, and furnished him with seed."[34]

An entry in La Verendrye's journal for the year May 27, 1733, to July 12, 1734, indicates some food problems, but also some harvest. "In this extreme need I made over to them (i.e. the Indians) the field of Indian corn which I had sown in the spring, and which was not yet entirely ripe. Our hired men also got what they could out of it. The savages thanked me greatly for the relief I had thus afforded them. The sowing of a bushel of peas after we had been eating them green for a long time gave us ten bushels, which I had sown the following spring with some corn."[35]

While this post was the farthest extension to the west in Ontario there were also some intermediate points at which some agricultural development was taking place. Thus, in 1761, Henry reported: "At the entrance to Lake Michigan, and at about twenty miles to the west of Fort Michilimackinac, is the village of L'Arbre Croche . . . The missionary resides on a farm, attached to the mission situated between the village and the fort, both of which are under his care. The Otawas of L'Arbre Croche, who, when compared with the Chipeways, appear to be much advanced in civilization, grow maize for the market of Michilimackinac, where this commodity is depended upon, for provisioning the canoes."[36]

"At the same time that I paid the price, which I have mentioned, for maize, I paid at the rate of a dollar per pound for the tallow or prepared fat to mix it. The meat itself was at the same price. The jesuit missionary killed an ox, which he sold by the quarter, taking the weight of the meat in beaver skins."[37]

Later, in 1769, he reported: "This year, I attempted to cultivate vegetables at Michipoten; but without success. It was not at this time believed, that the potatoes could thrive at Michilimackinac. At Michipoten, the small quantity of roots which I raised was destroyed by frost, the ensuing winter."[38]

At Grand Portage on Lake Superior, in what is now United States territory, there were some significant developments. Alexander Mackenzie gives a description of the situation when he visited there on his route west in 1788.

"The proprietors, clerks, guides, and interpreters, mess together, to the number of sometimes an hundred, at several tables, in one large hall, the provisions consisting of bread, salt pork, beef, hams, fish and venison, butter, peas, Indian corn, potatoes, tea, spirits, wine, etc. and plenty of milk, for which purpose several milch cows are constantly kept."[39]

That agricultural production at this post was important is evidenced by the fact that when the post was moved to the site at Fort William, in

1803, immediate steps were taken to provide for such production. Campbell writes: "Though as many as a thousand men had worked on its great hall, houses, stores, sheds and wharves at a cost of upward of £50,000, the depot was not yet completed for the first meeting in 1803. But already a few cattle grazed on the lush flats upstream, and on the rise of ground immediately above the post — where La Verendrye had camped half a century before — small fields were being tilled for seeding to coarse grains and potatoes."[40]

Another description stated: "Outside the fort is a shipyard, in which the company's vessels on the lake are built and repaired. The kitchen garden is well stocked, and there are extensive fields of Indian corn and potatoes. There are also several head of cattle, with sheep, hogs, poultry, etc., and a few horses for domestic use. The country about the fort is low, with a rich moist soil. The air is damp, owing to frequent rains, and the constant exhalation from Lake Superior."[41]

We see then that during the period prior to the American Revolution there had been numerous attempts at agricultural production in various parts of Ontario. However, at no place except in the area around Detroit had there been any real development and settlement, and even there it was limited to a few hundred people. In this respect the situation was quite different than that in Quebec at the same time. But each province offered some attraction for the Loyalists who were seeking a new home and each received a substantial contingent in the years following the peace of 1783. In fact some arrived before that time with the first lot taking up land at Niagara in 1780 when five families arrived there to settle. A survey in 1782 showed that there were 17 families with 49 children plus one hired man and one slave. In that year they had 236 acres of cleared land which had produced 206 bushels of wheat, 926 of Indian corn, and 46 of oats as well as 630 bushels of potatoes. They possessed 49 horses, 61 cattle, 30 sheep, and 103 hogs.[42]

This group was the forerunner of thousands that arrived after the peace. Also included amongst the new settlers after that time were discharged soldiers and still later emigrants from Britain. An indication of the dual movement immediately following the peace is seen in a return made in 1784 which showed that, above Cataraqui, 1,755 discharged troops and Loyalists took up land.

The Loyalists came mainly from pioneer areas of the United States and, therefore, were familiar with the process of clearing and preparing forest land for production. The land to which they came varied greatly in terms of climate, soil quality, and forest cover. A general description of the latter, and its relation to soil quality is given by Talbot when he notes:

"In every part of America, the quality of the soil is ascertained, more by the timber which it produces, than by the appearance of its surface or the nature of its substrata. Land, upon which black and white Walnut,

Chestnut, Hickory, and Basswood, grow, is esteemed the best on the continent. That which is covered with Maple, Beech, and Cherry, is reckoned as second-rate. Those parts which produce Oak, Elm, and Ash, are esteemed excellent wheat-land, but inferior for all other agricultural purposes. Pine, Hemlock, and Cedar land is hardly worth accepting as a present. It is however difficult to select any considerable tract of land, which does not embrace a great variety of wood; but, when a man perceives that Walnut, Chestnut, Hickory, Basswood, and Maple, are promiscuously scattered over his estate, he need not be at all apprehensive of having to cultivate unproductive soil."[43]

Other writers made somewhat the same observation but Shirreff[44] did not agree. He thought that the more direct approach of examining the soil with the aid of a spade was more exact and desirable.

Another factor affecting land desirability was the presence or absence of stones. Strickland[45] noted that the best land might be full of boulders that would be difficult and expensive to remove. He noted that a few boulders were not particularly objectionable, a few might even be an advantage as they could be used to construct French drains or could be rolled into the bottom of rail fences.

While most of the land was quite heavily wooded there were large areas of relatively open country with the trees widely dispersed. As a general statement it can be said that, with judicious choice, all of the early settlers could have obtained good quality land of relatively high fertility. That many of them settled on very poor land can be attributed to a number of factors including ignorance on their part and to improper information supplied by land agents and speculators.

The Loyalists and retired army personnel received generous government support in settling as they were given free grants of land and a full issue of rations for two years. They also received a stock of the primitive implements of the day, and other articles, including seed and some livestock, necessary for making a start in the wilderness. The government also provided grist mills which operated free of toll until 1791.[46]

Despite this support the first few years were years of hardship in getting land cleared and crop production started. It was even harder for some of the immigrants who came directly from Britain and lacked the knowledge and background of pioneer life. Land clearing and building shelter were the first requirements for all of the settlers and this was a slow process. Guillet has summarized the information on the rate at which land could be cleared:

"Various estimates are given as to how fast and at what cost the clearing of land could be accomplished. James Logan was told that an expert could chop ten to twelve acres in four months; but Samuel Strickland said that if the underbrushing had been done earlier a good workman could chop, cut, and pile an acre in eight days. Six men used to the work, said a guidebook, could chop and burn an acre a day, assuming, apparently, that the trees were dead. An American chopper, said another, could fell the trees on an acre in a week, but could not burn them. A survey made in Glengarry County showed that the average settler had cleared some 22 acres in his first three years. It would seem unlikely that a man working alone could clear, seed, and fence ten acres in a year, as a highly optimistic writer estimated; the average settler, in fact, would probably have been able to accomplish half as much. Of 11 pioneer settlers in Norwich Township, six were unable to sow any seed the first season, but the other five seeded from four to 14 acres each."[47]

Strickland emphasized the need for the settler to concentrate on cutting and clearing in the first three years because after that he would have to spend time attending to increased stock, barn and house building, thrashing, ploughing, etc. and would have less time for land clearing.[48]

In commenting on the cost of clearing, Shirreff, in 1835, noted: "The expense of clearing, fencing, and sowing depends on the nature of the timber, and varies from £3.10s to £5 per acre. The succeeding wheat crop, also, varies from 12 to 25 bushels per acre, and prices from 1s 6d to 5s per bushel. Generally speaking, money is not rapidly made by clearing forest land, while patient industry seldom fails of being ultimately remunerated."[49]

At another point he indicated the expense of obtaining the first wheat crop as:

"Purchase money of wood land	$ 3
Under brushing and chopping	8
Logging, burning, fencing	8
Seed and harrowing	3
Carting and harvesting	2
Thrashing and teaming	5
	$29

Produce estimated at 25 bushels, at $1 per bus. = $25."[50]

At Whitby he was informed by one of the farmers that "He had let twenty acres of forest land to clear and fence, at $12 per acre, which he says is the common cash price of the country."[51]

According to Howison[52] waste land could be cleared and fenced at £4 per acre, though a lower cost could be obtained for large areas under contract terms. Daily wages for farm labourers ranged from three shillings to four and sixpence, exclusive of board. On a monthly basis they were £3 per month besides board, though the rate would be lower if the man was hired for the year.

In land clearing the stumps were not removed until later years when they had rotted. The first crops were seeded or planted between the stumps, in most cases without any land preparation other than the removal of the trees and the underbrush. Grain crops, usually wheat at first, to provide flour for home use, were seeded between the stumps and the ground was then stirred with a harrow to help cover the seeds. The standard harrow was triangular in shape, made of heavy timbers through which iron teeth were driven. The poorer settlers even managed without a harrow or oxen to pull it as we see from a letter written in 1832:

"You will no doubt wonder how we commence farming without money, we have no want of anything but an axe, hoe, and rake; we cut down the trees and burn them, rake the ground, and burn the stuff, then sow our wheat and hoe it in the same as gardiners do their small seeds in England, and this mode of culture will do most soils for three years and will produce from thirty to forty bushels per acre of wheat."[53] This estimate of yield is considerably higher than the estimates given by most observers.

The system of planting potatoes was quite simple, though details varied slightly. One description was: "Potatoes on new land are also planted with the hoe, and in hills about five thousand to the acre. A hole is scraped with the hoe, in which four or five sets, or a whole potatoe is dropped. The earth is then heaped over them in the form of a mole-hill, but somewhat larger. After the plants have appeared above the surface, a little more mould is drawn around them. Very large crops of potatoes are raised in this manner. Two hundred and fifty bushels per acre are no uncommon crop. I have assisted in raising double that quantity; but of late years, since the disease has been prevalent, but poor crops have been realized."[54]

For Indian corn the system was somewhat similar, but with the additional feature of intercropping. In this case the corn was planted in rows three feet apart and with the hills 30 inches apart in the row. A pumpkin seed or two were sown in every second or third hill in the row. This system provided for two crops in the year. At a later stage of land development a visitor noted that in many cases wheat was sown between the hills of corn. This permitted the wheat to be well advanced when the corn was harvested.[55]

One general practice was to seed grass seed, usually timothy, and clover with the wheat so that a grass or hay crop could be obtained in the second and third crop year. Then newly cleared land would be used for grain, potato, or corn production.[56] It would usually take up to seven years for the stumps and roots of the hardwood trees to rot sufficiently for them to be removed without too much difficulty and make the use of the plough feasible. There was little that the settlers could do to modify this system and, as one observer noted, "There is little room for the display of genius or management, the process being nearly the same in all cases."[57]

In the longer run the gradations in development of the land and the practices used in production appear to have been very great, from the time of first settlement to the middle of the 19th century. They ranged from the most primitive to a relatively progressive system of production. Not only were there real differences, but there also were differences in the evaluation made by the various observers who left accounts of their experiences and/or observations. Keeping in mind that the first influx of settlers began in 1784, it is interesting to note some of the early reports. Thus, Campbell, visiting North America in 1791 and 1792 wrote of the Niagara area: "Wheat is rarely left here above a day on the ground after reaping, and often carried home to the barn the very day it is cut; the ground is no sooner cleared of one crop, than it may be, and often is, immediately plowed down and sown with another, and so on alternately without using any sort of manure. The richness of the soil and the salubrity of the air, make all sort of stimulus totally unnecessary."[58]

In this connection it should be noted that in Ontario winter wheat was preferred to spring wheat, in contrast to Quebec where winter wheat had been tried and found to be less successful than the spring-sown types. Climate accounts for the difference between the two regions. As Father Hennepin had noted, there was a difference of three months in the length of winters between Quebec City and Fort Frontenac.

Another observer, in a letter to a friend in England in 1794, noted regarding the Niagara area: "In many places there is little more for the farmer to do, than cut a sufficiency of timber to fence his fields, girdle or ring the remainder, and put in the harrow, for in few places only is it necessary to make use of the plough, till the second or third crop, there being little or no underbrush."[59] Among other things he added: "Orchards are in great forwardness, for the age of the settlement, some of which already bear fruit. Peaches, Cherries, and currants are plenty among the first settlers. The farmers raise a great quantity of pork, without any other expense than a little Indian corn, for a few weeks previous to killing, and often kill their hogs out of the woods, well fatted on nuts."[60]

He was particularly impressed with the Bay of Quinte area where he noted that the soil was rich and easily worked. It produced from one to three crops with only harrow cultivation and yielded from 20 to 30 bushels of wheat per acre.[61]

Weld, who travelled in 1795, 1796, and 1797, wrote of the Malden area: "Amongst the scattered houses at the lower end of the district of Malden, there are several of a respectable appearance, and the farms adjoining them are very considerable. The farm belonging to our friend, Captain E -, under whose roof we tarry, contains no less than two thousand acres. A very large part of it is cleared and it is cultivated in a style which would not be thought meanly of even in England."[62]

If this was true it would be the exception rather than the rule, as will be seen from later comments. Weld went on to say: "Beyond Malden no houses are to be seen on either side of the river, except indeed the few miserable little huts in the Indian villages, until you come within four miles or thereabouts of Detroit. Here the settlements are very numerous on both sides, but particularly on that belonging to the British. The country abounds with peach, apple, and cherry orchards, the richest I ever beheld; in many of them the trees, loaded with large apples of various dyes, appeared bent down into the very water. They have many different sorts of excellent apples in this part of the country, but there is one far superior to all the rest and which is held in great estimation, called the pomme caille. I do not recollect to have seen it in any other part of the world, though doubtless it is not peculiar to this neighbourhood. It is of an extraordinary size, and deep red in colour, not confined merely to the skin, but extending to the very core of the apple; if the skin be taken off delicately, the fruit appears nearly as red as when entire."[63]

Heriot, in 1806, noted: "The exuberance of the soil around the Bay of Quinte, amply rewards the toils of the farmer; it is worked with facility, and produces many crops, without application of manure. The usual produce is twenty-five bushels of wheat per acre."[64] In referring to the Niagara area he wrote: "Families from the United States are daily coming into the province, bringing with them their stock and utensils of husbandry, in order to establish themselves on new lands, invited by the exuberance of the soil, the mildness of the government, and an almost total exemption from taxes. These people either purchase lands from the British subjects, to whom they have been granted, or take them up on lease, paying the rent by a portion of the produce."[65]

This latter statement indicates that a new pattern was developing with continuing immigration, the beginning of trade in land and a second generation of owners taking over. In the earlier statements can be seen the source of future problems, similar to those encountered in Quebec, in the continued cropping of land without any attention being given to maintaining the fertility of the soil. The early fertility was there but, in most cases, it was insufficient to sustain a high level of production for an extended period of time.

One of the basic problems facing the early settlers was the shortage of feed for livestock, especially for winter feeding. It was not until fields of tame grasses and clovers were established that there was a proper feed supply available, though corn stalks, straw, and pumpkins partly filled the gap.

Jones draws a grim picture of the situation: "But necessary though oxen and milch cows were to him, the backwoodsman accorded them the worst of treatment conceivable. He worked his oxen from early morning to nightfall without food, and then turned them into the woods to browse. He milked his cows at strange hours, and sometimes not at all for days. Fortunately cattle managed to thrive on browse and other coarse feed during the summer. He often provided no shelter for them when winter came, and after the pumpkins were all consumed, gave them nothing to eat but straw. It was estimated that 1,500 cattle perished in London Township, Middlesex County, in 1822 from poor feed and lack of shelter, and corresponding numbers in all the adjacent townships."[66]

Strickland portrayed another aspect of the problem. He maintained that there was a plentiful supply of summer feed in the woods, beaver meadows, and the margins of lakes and streams so that cattle were in good condition in the fall of the year. However, the trouble was the time it took to find the cattle that were turned out into the woods. As he noted: "I have myself often spent two or three days in succession, searching the woods in vain; and it not infrequently happens that, while looking for the strayed beasts, you lose yourself in the woods."[67]

Trail[68] had a partial solution to the winter feed problem, but one that could be used only in certain circumstances. The solution was to sell the cattle in the fall and buy others in the spring even though this appeared to result in a loss; however, it might be less than losing the cattle altogether during the winter.

In the chapter on Quebec considerable attention was given to the poor level of husbandry practised by the French farmers as contrasted to the apparently more progressive English. It now appears that the English in Ontario were, in fact, no better in many respects. Dunlop, writing about 1832, noted:

"Of Agriculture, as practised in this province, I have very little to say, except that were the same slovenly system pursued in any country less favoured by nature, it would not pay for the seed that is used. I have already stated the ruinous mode of taking repeated crops of wheat off the land; and on the river Thames, in the Western district, I witnessed a refinement of this barbarism, viz., burning the stubble before the land was ploughed for winter wheat, and thus depriving it of even that trifling strength that it might derive from the decomposition of the straw."

"It is only in some parts of the province that manure is used at all and it is not an uncommon occurrence, when the stable litter has accumulated in front of the building called the barn, (which generally contains all the farm offices) to such a degree as to become a nuisance, that a man invites his neighbours to assist in removing the barn, which is always a frame building, away from the dunghill, instead of transporting the dunghill to the wheat field."[69] These sentiments have a familiar ring as similar criticisms had been levelled at the farmers in Quebec and the Atlantic provinces.

Some years earlier Talbot[70] had summarized the situation in detail, claiming that the average production was no more than two-thirds what it could be under proper management. He emphasized the lack of use of manure, and the absence of crop rotations and summerfallow. The end result was land exhausted of fertility.

Not all observers took quite so dim a view of the situation for Howison noted: "Between Queenston and the head of Lake Ontario, the farms are in a high state of cultivation, and their possessors are comparatively wealthy. Some of them contain more than one hundred and fifty acres of cleared land, the fields of which have become smooth and level from frequent ploughing, and are not disfigured by stumps and decaying timber. A great majority of the individuals who are owners of these farms, came to the Province twenty or thirty years ago in the character of needy adventurers, and either received the then unimproved land from the government, or purchased it for a trifle... Many of them possess thirty or forty head of cattle and annually store up to two or three thousand bushels of grain in their barns."[71]

No doubt there was a great range in the level of management, but on the whole improved cultural practices had not become general. Various factors were involved such as the origin and background of the settlers. Thus, the German, Dutch and Low Country Scots were recognized as being the most advanced in terms of the cultural practices used and the care that they gave to their crops and their livestock. On the other hand many of the settlers came from what today would be classed as the underprivileged. Financially they were poor when they arrived, had relatively little education and had no agricultural experience. Their main endeavor was to survive and there was no strong incentive for them to introduce major changes.

Another factor was that land continued to be relatively cheap and labour relatively expensive and in short supply. Under these conditions it was a question of how much effort and expense should be given to maintaining soil productivity.

Marketing a Challenge

There was also a lack of markets for the farm products and especially for wheat and flour which became the main export products. During the early years, the first comers could sell their produce to the military garrisons and to the burgeoning settlements. But as production increased, larger markets were needed and these were found mainly in Montreal and Great Britain. The availability of these markets fluctuated widely. The Montreal market *per se* depended primarily on local demand which varied with the ups and downs of crop production in Quebec. Access to the British market was controlled directly by the Corn Laws of Britain which set prices for imports. The success or failure of the British crops at any time affected the price and, thus, indirectly determined when the market would be open to Canadian wheat at a favourable price.

A somewhat special situation developed in the Ottawa valley with the burgeoning lumbering industry of the early 1900s providing an outlet for practically all agricultural products. This industry required hay and oats to feed the many horses and oxen used in lumbering, and it required flour, vegetables, and livestock products to feed the men engaged in the woods. In addition it provided employment for the settlers and for their horses and oxen during the winter when they were not needed for farming. Unfortunately, these opportunities were accompanied by adverse effects. Too often the settlers became so involved in the lumbering activities that their farming activities were partially neglected and the productivity of the farms was not maintained.

A period of prosperity occurred during the war of 1812 with the United States as the military absorbed all available products. Production in Ontario was inadequate to supply the demand despite an embargo on export of wheat during these years. When peace returned prices dropped drastically with the usual depressing effects on farmers.

A critical limiting factor in marketing was the inadequacy of transportation. The immediate problem was the lack of roads for local transport, but more significant was the fact that the only practical avenue for export was by way of the rivers and lakes. The St. Lawrence river was the main route to Montreal and on to Britain, but it presented obstacles in the form of rapids that were difficult to navigate and limited the type of vessels that could be used. Above Lake Ontario the main obstacle was Niagara Falls which was not overcome until the completion of the Welland Canal in 1829. The significant role of the waterways led to the development of the main markets for wheat at ports on the lakes. Local merchants and millers at inland centres provided a secondary market, but many farmers hauled their wheat as much as 100 miles to the ports in order to get the higher prices usually available there.

Wheat was the major export crop because it had a broad demand, was best able to cover transportation costs, and was relatively easy to handle for transportation. Corn was not grown widely as only in the more favoured areas in the southwest corner of the province would it ripen in most years. Other crops included potatoes, rye, oats, barley, and millet in about that order of importance.

Talbot,[72] in 1823, gave a detailed description of crop and livestock production practices which provides the basis for the following summary.

Corn required four quarts of seed per acre and produced an average of 25 bushels per acre. Sometimes it was planted in drills, with pumpkins interspersed. These could yield up to 1,200 pumpkins per acre and provided food for the settler and feed for his cattle.

Winter wheat was generally sown between the first of August and the middle of September. Spring wheat was seeded about the 20th of April and was ready for harvest by the end of August, or about a month after the winter wheat was harvested. The average yield of winter wheat was about 25 bushels per acre whereas spring wheat yielded somewhat less and was considered to be not so safe a crop. Seeding rates ranged from 45 to 60 pounds per acre.

Rye was grown quite widely, mainly for the distilleries. The average yield was about 25 bushels per acre and the price about 25 per cent lower than wheat. Oats were considered to be a most unprofitable crop of very low quality. Barley was not a particularly important crop with average yields of about 20 bushels per acre.[73]

Millet was cultivated in some areas. Seeded at the rate of three quarts per acre it would produce 80 bushels per acre on good land. Forage crops included White Clover, which sprang up spontaneously when the land was cleared, Timothy, Red Clover, Lucerne, and Herd-grass.

Potatoes were an important crop with an average yield of about 175 bushels per acre though Talbot was confident that, if the Irish system of planting in drills was used, the yield could be quadrupled.[74]

In general the management of livestock was very bad and the quality of the animals was poor. Cattle were described as being at least a third smaller than those of Great Britain and Ireland. This was ascribed to the fact that they were never housed in winter and were poorly fed. "They are seen in the severest weather, when the snow is almost deep enough to cover them, skulking around the barn doors; where, one would think, their pitiful looks and sunken sides would be sufficient to extort provender from a heart of stone. Notwithstanding the inhuman treatment which they receive during the Winter, they are found in excellent condition soon after the return of Summer, and give, I believe, nearly as much milk as the best English cows."[75]

Sheep were described as miserable-looking, seldom weighing more than 50 pounds per carcass and yielding only about two and one half pounds per fleece. On the other hand, hogs were described as being excellent, exceedingly hardy, though not large. they were usually killed when a year and a half old when they would weigh about 200-weight.[76]

Though the general level of agricultural practices left much to be desired there were some early moves made to improve the situation, first organized by agricultural societies. One of the first such societies in Canada, and certainly the first in Ontario, was formed at Niagara either in 1791 or 1793. It had the support of Governor Simcoe who stated to the society that: "while he shall continue in the administration of this Province he intends to send annually Ten Guineas to be disposed of in premiums for the improvement of agriculture." There is no indication that this society lasted very long nor is there any record of any specific action by the society for the benefit of agriculture.

Another society, the Upper Canada Agricultural and Commercial Society was formed in 1806, at York,[77] but there is nothing to indicate that any action was taken. This society did not last long and apparently ran into financial difficulties as can be seen from a report in 1807. "At the meeting first above mentioned, it was discovered that the Funds of the Society had found their way into the hands of the most able and upright President, who in the multiplicity of momentous concerns in which he was engaged, forgot to remind the society where its cash was deposited. It is characteristic of a truly great man (such as he was) to be superior to little punctilios or minutiae in money matters: the sublimity of his genius, soaring above the groveling level of ordinary mortals, descends not to consider twenty pounds as a subject worth mentioning. This desecration of our treasury we should have supposed connected with another patriotic act, namely, the liquidation of his election accounts, did not the lamentations of Martin Holder & Co. destroy the supposition."[78] In any event, at a meeting on February 6, 1808, the society was dissolved, but the members present were not ready to call it quits for they decided to hold another meeting a week later for the express purpose of forming a new society.[79] There appears to be no record of ensuing action.

It was not until the end of the first quarter century that new societies were formed, but then considerable activity developed, especially after 1830 when the legislature passed an act providing for financial support to such societies.

The most visible activity of the agricultural societies was the staging of fairs and plowing matches. Robinson states that an agricultural fair was held at Queenston in 1822, claiming, incorrectly, that it was the first fair in Canada.[80] Guillet indicates an even earlier fair stating: "An agricultural fair is known to have been held at Queenston before 1800."[81] Be that as it may, numerous local and district fairs were instituted and became an integral part of agricultural life.

An indication of the scope and focus of these fairs can be seen from the following two examples. The Cumberland County Agricultural Society was organized in 1828 and held a fair on October 19, 1829, with first and second prizes for the following classes: best stallion, best bull, best brood mare, best milch cow, best pair oxen, best boar, best ram, best plain roller, two best cheeses, and best written essays on the culture of wheat. It was noted that: "The two essays were read publicly to several hundred persons."[82]

The Home District Society held a fair on May 20, 1833, with first, second, and third prizes for: stallions, mares, cows, working oxen, working horses, and samples of three bushels of oats, barley, peas, and potatoes. In addition: "£50 to be distributed amongst such persons as shall grow on one acre of ground the largest quantity of the best Wheat, Barley, Oats, Peas, Potatoes and Indian corn."[83] This fair was followed by one on October 17 of the same year with first, second, and third prizes for: bull, ram, six ewes, sow, ram lamb, ewe lamb, spring colt, and spring calf. A ploughing match was staged with the following prize list:

		£	s	p
First prize	An Iron Plough value	7	10	0
Second	A Plough Harness	4	0	0
Third	Pair of Brake Harrows	3	0	0
Fourth	Pair of Seed Harrows	2	10	0
Fifth	A Drill Harrow	1	10	0
Sixth	A Scuffler	1	10	0 [84]

The most noticeable feature of the prize lists for livestock is that they made no reference to breeds, indicating that breeds, as such, were not receiving much attention.

However, the societies did become involved in importing livestock though the benefit from this might be questioned if the Home District Society action was a fair sample of this activity. It ordered that £100 should be appropriated for the purchase of bulls in the United States, hoping to get three full blood or two full blood and two three-quarters blood bulls for this amount of money. They were to be let at public auction for one year, at the fair to be held on May 20.[85] The success of this action may be judged by the statement made by a visitor to the fair: "There was a fair and cattle show here yesterday (May 20), a poor concern. The bulls, inferior brutes, were lett for the year to members of the Agricultural Society for from £5 to £7 10. The number of stallions was considerable, yet with the exception of 1 or 2, they were poor sticks."[86]

It is difficult to get a picture of the extent of livestock importation for improvement purposes, but it is clear that a substantial amount took place, both from the United States and Britain. For example, in 1835, Shirreff wrote: "While at Kingston, ten shorthorned cattle, nineteen Southdown sheep, and a lot of swine, came to the yard of the Kingston hotel, at which I lodged, on their way to the county of Dumfries, Upper Canada, direct from England."[87]

To begin with, all of the agricultural societies were separate and distinct organizations with no coordination among them. In 1846 the Agricultural Association of Upper Canada was formed to act as a coordinating body. One of its first actions was to organize the first provincial fair at Toronto that year at which premiums of £225 were offered, about £200 being contributed by district societies.

Newspapers of the day also may be considered to have made a contribution to agricultural improvement by publishing articles and letters on agricultural subjects. There was quite a variation among the papers, as measured by the number of agricultural items that appeared. Basically, the content of the articles was the same in all of them with most of it being cribbed from American and British papers. These items provide good examples of the information available to farmers at that time and some of it provides very interesting reading. It is difficult to appreciate fully, in light of current knowledge, the limited amount of good practical information available at that time, to say nothing of the lack of scientific knowledge bearing on agriculture.

A contribution from a local writer appeared in 1796, castigating the farmers for their poor husbandry practices. In 1797 the *Upper Canada Gazette* invited contributions and over time received quite a number. A sampling of subjects that appeared in the various papers includes: how to prevent canker worms from destroying apple trees; account of a method of preventing decay of peach trees; mode of breaking steers to draft in a few days; sheep improvement; a recent invention for renewing the vigour of fruit trees; culture of potatoes; use of turnips as a green manure crop; Egyptian wheat; control of insects in the garden; a recommendation for the use of ashes on corn land; on raising turkeys; grafting apple trees; culture of flax; a certain cure for measles in swine; watering horses; experiments with carrots; on giving salt to cattle.

Three items will be quoted as they contain very interesting information and two of them illustrate clearly the empirical, almost witchcraft, approach underlying the recommendations.

The first one is: "A cure for yellow water and bots in horses. Take one pound of Roll Brimstone, two ounces of Rosin, two ounces of Antimony, two ounces of Aloes, two ounces of Salt Petre, one thigh bone of a stud horse, burnt, pulverize the whole to powder, then give the horse that has the yellow water or bots a portion once in three days till you see the symptoms change. It is also useful to give horses that are well, to prevent them from catching the disease. A tablespoonful is sufficient for a portion and it must be given in their grain, remembering to keep them from cold water about twelve hours, and let the first be given warm and with bran. We recommend this to farmers who have ailing horses in the Spring. Jabez D. Hammond and Isaac Seeleyand."[88]

The second item is: "An account of the new insects so prejudicial to apple trees, and a method of extirpating them. To one hundred gallons of human urine add one bushel of lime; add cow dung to bring it to the consistency of paint; with this composition anoint the trees. The present is the proper season for applying it. If the white efflorescence-like substance in which the insects are lodged has made its appearance, it should previously be brushed off."[89]

The third item is somewhat more conventional and deals with the fattening of hogs. "The fattening of winter hogs is a matter of importance, and by proper management, much may be saved. As soon as the Indian corn is fit to feed, the hogs should be put in a pen. Rye, buckwheat, or corn, boiled together; boiled potatoes, pumpkins etc. make excellent slop for them, it should be made thick, and given three times a day; and as much of the soft Indian corn as they will eat clean."

"Hogs should be fed a little at a time, but often and a great care taken not to stall them. The last two weeks of feeding them, Indian corn and water are the best, their pen should be cleaned twice a week and their bed made of clean straw. Swine are liable to a variety of diseases, to wit: mange, measles, stopping of the fore legs, etc. These diseases it is believed, proceed from the filthy manner in which hogs are generally kept. The mange may be cured by sprinkling on the backs of hogs, woodashes, and letting them out on a rainy day, after putting oil on them; a more efficient way is to wash them in soap suds, and oil them; after this trouble keep them clean. For measles, the flour of brimstone is said to be good. For the stoppage of the issue of the forelegs nothing more is necessary than to open the holes with the end of a knitting needle, or something of that kind."

"Boiled apples, peas, pumpkins, potatoes and squashes, all make excellent food for young shoats, and much the cheapest. Every farmer who wishes to raise pork for the market ought to have a boiler fixed for that purpose. After harvest, hogs should be turned into the orchard, that they may get the early ripe apples, which fall before the season of cider making commences."[90]

Government Support Programs

Colonial government policy, in Ontario as in the other eastern provinces, was to encourage the production of hemp for the British market. One form of encouragement was financial, and acts providing for funds were enacted from time to time. The newspapers also gave support to this program and published articles dealing with production practices. In 1800 "The Honorable the Trustees for the Encouragement of Arts, Manufactures, and Commerce," also entered the picture with awards for the production of hemp. These were a gold medal to Mr. J.W. Clarke of Montreal, a gold medal or one hundred dollars to Mr. J. Schneider of York, and a silver medal or eighty dollars to Mr. D. Mosher of Kingston.[91] Even the Board of Agriculture in London, England, got into the act with a reward of fifty guineas for the best report to the Board on the present state of cultivation of hemp. But all of these efforts were of little avail. The farmers of the province, like their counterparts in the other eastern provinces, never gave this crop a prominent place in their production programs.

More success was achieved with tobacco despite the fact that this crop did not receive encouragement from the government. The crop was introduced some time before 1821 for that year the *Weekly Post* recorded: "Experience has proved that the culture of Tobacco may be successfully and very profitably carried on in this Province."[92] Actually, tobacco had been raised by the Indians prior to the coming of Europeans. The rate of increase in the production of this crop is not known but the crop first appeared in the census records in 1851 with a production of 764,476 pounds, which would indicate considerable acreage.

Data portraying the growth of agriculture up to the middle of the 19th century are shown in Table I. The first entries show only population data as no agricultural data were collected. However, practically all of the early settlers were engaged in agriculture; thus, increases in population meant increases in agriculture even though the acreages might be small for each settler. The second quarter of the century witnessed rapid expansion in acreages cultivated and in the production of various items.

The early farmers first of all had to conquer the forest and get land into production, but after that they were faced with innumerable hazards to their crops and livestock without adequate means of controlling or overcoming them. The weather caused problems at times. In 1829 the crop was reported to be seriously affected by drought and the same report stated that the wheat crop of the previous year had been a failure though the reason was not given.[93] Conversely, in 1831, the wheat crop did not mature and as a consequence the flour made from it soured and the "cargoes of wheat exported to London and Liverpool became solid masses in the ships and had to be dug out with shovels."[94] Shirreff noted that among the samples of wheat that he saw in a mill at Newmarket, in 1834, many were sprouted.[95] It is quite likely that a crop failure occurred in 1804 for on May 11, 1805, the baker in York placed the following notice in the *Gazette*: "To the public: On account of the present scarcity of flour, the Subscriber is under the disagreeable necessity of raising his bread to Fifteen pence, New York currency per loaf, not being able to afford it for less after this date. (May 3, 1805)."[96] Later in the year another notice appeared raising the price to 18 pence per loaf.

Early frost was a hazard with damage to crops leading, in some cases, to secondary problems as this notice in 1817 would indicate. "To Farmers — We state for the information of those who are in the habit of feeding their Cattle, with poor or injured corn, which did not come to maturity this year, owing to the early frosts, that several valuable Cows and many young cattle have died in this town the past week, which had been fed on this kind of food, and the milk and cream of others which survived, is so offensive as to be unfit for use."[97]

Insects of various kinds, including mosquitoes, black flies, and other pests that attacked both humans and livestock were also a problem. However, of greater damage to production were the insects affecting crops. Grasshoppers were important though possibly not quite so frightful as described by Talbot who claimed: "Some of the Grasshoppers in these Provinces are as large as a field-mouse, and all of them are much larger than any I ever saw in Europe."[98]

Shirreff, in 1835, mentioned that "a dark-coloured caterpillar had devoured some fields of timothy grass, with the exception of the culms, and the insect had extended its ravages partially to Indian corn and wheat, but red clover was untouched, growing amongst the timothy which had been consumed." He referred to the fact that this insect also had caused damage in 1825.[99]

In the chapter on Quebec, reference was made to the ravages of the Hessian fly and the wheat midge. Hind stated that the Hessian fly was prevalent as early as 1805 in Lower Canada, but did not appear in Upper Canada until 1846 or 1847. However, Jones, after recording an export of 12,823 bushels of wheat, 896 barrels of flour, and 83 barrels of middlings from Kingston in 1794 stated: "The incipient export trade was checked by the prevalence of the Hessian fly, which assisted by dry summers, reduced the production of wheat for a number of years beginning with the harvest of 1794."[100] The wheat midge made its appearance somewhat later, Hind referring to it in Lower Canada as early as 1828, but not in Upper Canada until 1849 and 1850.[101]

In their efforts to overcome the ravages of these pests, and also diseases, the farmers introduced new varieties of wheat from time to time. Most emphasis was given to winter wheat in the hope of planting varieties that would mature early enough to escape severe damage by the insects or by rust. Some success was achieved though complete control was not achieved. In the areas where spring wheat was grown similar action was taken and numerous varieties were introduced, including Red Fife which became the most successful variety.

The turnip crop also was damaged by an insect, the fly. An interesting proposal for the control of this insect appeared in the *Gazette* on November 2, 1805: "A very important secret was made known for the communication of which 200 guineas was previously subscribed. It is a preventative against the insect called fly. The discovery is to sow 2 lbs. of radish seed on every acre of turnip land, with the turnips, which the inventor declares will so attract the fly, as to prevent its proving at all injurious to the turnips."[102] This activity could be classed as an early example of biological control, if it worked. In any event, it must have been considered to be important in view of the prize that was given for it.

Of insects affecting animals, other than mosquitoes and black flies, Heriot refers to ticks on an island opposite Kingston and states: "To horses and cattle, which have been sent to graze, on this island, the ticks, from their multitudes, have been frequently fatal."[103]

Two plant diseases were of particular concern, namely, smut and mildew, the latter possibly being synonymous with rust. Smut was generally present and troublesome, not only for the damage that it caused to yield, but because it could not be removed from the grain and, consequently, interfered with the production of high-quality flour. It may well have come to the province with the first wheat. It was referred to specifically as early as 1810 and Reaman suggests 1800 as the time of its first appearance.[104]

Mildew was possibly the most detrimental disease of wheat. It can not be determined when it first made its appearance, but certainly it was early in the agricultural history of the province. Shirreff refers to it as damaging crops in widely scattered parts of the province in 1834. Jones refers to black stem rust as being particularly destructive in 1837, 1839 and 1842, and more or less so in every other year from 1840 to 1846 and again in 1849 and 1850.[105]

Disease also affected the potato crop, a complete failure being reported in 1844 "because of dry rot, hitherto unknown here."[106] This would be the blight that had struck the other eastern provinces slightly earlier.

Another pest, still with us but not especially troublesome to the farmer today, was the squirrel, for Strickland said: "There is another animal, which I think is more numerous than formerly; I mean the black squirrel. These pretty little creatures are very destructive amongst the Indian-corn crops. I have seen them carrying off a whole cob of corn at once, which I will be bound to say was quite as heavey as themselves."[107]

More troublesome, among the wild animals, were the bears and wolves which caused much damage to the domestic livestock, especially the sheep.

Weeds made an early appearance and, with the cultural methods and equipment generally in use, were not kept well under control. Talbot made specific reference to what he called "that indigenous weed, the Canada thistle", in 1823.[108] By 1849 there was no dearth of weeds and the *Canadian Agriculturist* listed no less than 34 annuals and biennials as well as 23 perennials. Among the latter were such well known culprits as the Canada thistle, buttercup, ox-eye daisy, dandelion, yarrow, golden rod, and couch grass.[109]

Throughout the early period of settlement, farming equipment was quite primitive and a limiting factor in expanding acreage under cultivation on individual farms. At first the axe, spade, pickaxe, and hoe were the main cultivating tools and the sickle and flail the main harvesting tools. The harrow was the first animal-powered implement, a crude, A-shaped device, made of timber with iron teeth driven through the timber. It could be navigated between the stumps left after clearing the trees, and served to cover the seeded grain.

Though the plow could not be used effectively in most areas for the first few years after clearing because of stumps and roots near the surface, it made a relatively early appearance. In his address to the 1849 annual meeting of the Upper Canada Agricultural Association, the president reminisced about the early days of the Loyalists in the Bay of Quinte area. He stated that the old English plow was among the items provided to the Loyalist immigrants. About 1808 the "hog plough" made its appearance with "full iron share forming the front or rising part of the mould-board, the residue of which was still obliged to be made of wood.... About the year 1815 the farmers generally fixed their attention upon the cast-iron share and mouldboard, all cast in one piece."[110] Gradually improved models came into use, coming mainly from the United States and Britain. One popular model was the Scotch swing plow.

For harvesting, the sickle was replaced by the scythe and the cradle. This was a big step forward as the harvesting could be done more rapidly, but it was hard work. A description of harvest with the cradle is given by Logan:

"The wheat being ripe, he commenced cutting on the 24th August, when two neighbours came with cradles. We all turned out at six o'clock, A.M., there being three cradlers and three binders. My brother's two bondsmen, one of whom was an Englishman, could not keep up with their cradlers, so that he was obliged to assist them in binding; but the other binder being more active, and having been several years in Canada, kept pace with his cradler. The large thistles, some of them seven feet high, cause us great annoyance. We went home to breakfast at eight, returned in an hour, worked until one, when we had dinner, resumed our labour at two, and continued until six. It is customary to give every two men a bottle of whiskey to mix with water... The three cradlers cut five acres daily, so that the eighteen acres of wheat were all reaped in less than four days. Much corn is wasted, both because the cradlers scatter it about too much, and because the binder is too much hurried to keep up with him. What is left in the field, however, is used for fattening pigs."[111]

It is no wonder that the mechanized reaper, first brought to Ontario in 1843, found favour and came into general use. With the reaper one could harvest 10 to 15 acres per day, but it still required considerable labour to bind the grain into sheaves. For example, Jones[112] stated that with the Hussey machines 11 men were required, one to drive the horses, one to rake the sheaves off the platform, and nine men to bind the sheaves. But even so the productivity per man was considerably greater than under the old system.

A mechanical threshing machine was introduced earlier than the reaper, the first being brought to Ontario in 1832, and this was soon followed by others. It is of more than passing interest that a combine harvester was under development even earlier than this. In April 1818 the announcement of such a machine appeared in the *Gazette*: "Patent harvester. The model of a machine to cut, thresh and clean wheat, rye, oats, barley, etc. at one operation, is now exhibiting at the Tontine Coffee House, by the inventor and has attracted the attention of many men of science, and practical agriculturists, all of whom we understand express the highest opinions of its merits; the machine is constructed to be moved by the strength of one horse — enters a field of wheat, rye, etc., will take 'a two man's land ahead' and cut, thresh, and fan the grain fit for the mill or market, and without waste or without leaving anything behind to be gleaned. This complete operation can be performed as fast as a horse can walk. The machine may be separated and used only for gathering the grain, which will render it extremely simple and effective. It is calculated that two horses, and a man to attend them will cut and gather from twenty-five acres per day. The net cost of a machine for cutting and gathering the grain will not exceed one hundred dollars; and a machine complete, for performing the operation of preparing the grain for the mill will about double that sum."[113] There is no record to indicate that a machine of this description actually was made and used in Ontario, but it is rather doubtful. It is interesting that the concept of a combine harvester existed at such an early date.

Legislation Enacted

The government of Ontario took early action to enact laws, regulations, and support for agriculture. The following list, while not exhaustive, gives an indication of the areas of concern to the legislators of the province.[114]

1792
- An act to establish the Winchester measure and a standard for weights and measures.
- An act to regulate the toll to be taken by mills.

1793
- An act to encourage the destruction of wolves and bears. This act was re-enacted and modified on several occasions.

1794
- An act to restrain the custom of letting animals run at large. This act was amended on several occasions and later included a pound law.

1797
- A land titles act.

1801
- In the first recorded action for the encouragement of the production of hemp, the Assembly prayed "The Lieutenant Governor to apply (from unappropriated funds): £250 for hemp seed to be distributed gratis and up to £500 for premiums and bounties."
- An act to appoint inspectors of flour, pot and pearl ashes.

1804
- An act to encourage the growing of hemp. This act was re-enacted and modified from time to time for many years, and substantial sums of money were expended.

1805
- An act to regulate the curing, packing, and inspection of beef and pork.

1814
- An act for the establishing and regulating a market at York.

1821
- A special Legislative Committee reported on problems of transpor-

tation for agricultural products, especially for western Ontario; on the need for inspection of flour to ensure quality; and on the need for ready access to the British market for wheat.

1822
- A new act for the regulation of the running at large of stock.
- An act granting money specifically for the purchase and erection of machinery for the preparation of hemp for export.

1830
- An act to encourage the establishment of Agricultural Societies by granting £100 for each society on the condition that each society subscribe half that amount. The resolution preceding this act is of sufficient interest to warrant recording:

"Resolved, that the experience of England and other enlightened communities, having shewn that the institution of Agricultural Societies in different sections of the country has greatly promoted the success of the farmer, it is wise and politic in this Legislature to give all the encouragement in its power to such Agricultural Societies as are, or may be founded on a liberal footing in this Province."

"Resolved, that if any appropriation, not exceeding one hundred pounds, in aid of such societies, was made for each and every district in this Province, it would materially assist, and induce the importation of valuable livestock, grain, grass seeds, useful implements, etc., but in no case shall such appropriation be made until the society applying for it shall make it clear that they themselves subscribed and paid into the hands of a regularly appointed treasurer, a sum equal to fifty pounds."

1835
- An act to establish a standard weight for the different kinds of grain and pulse.

1850
- An act to establish a Board of Agriculture. This was the beginning of a Department of Agriculture.

By the middle of the 19th century, agriculture had passed through the pioneer period in most parts of southern Ontario and was a well established industry. Most of the area had been settled, though not all of it had been cleared and brought into production. Already the population pressure was beginning to be felt, which, a quarter of a century later would lead to considerable emigration from Ontario to the lands in the West ideally suited for agriculture. On the better agricultural lands in Ontario a pattern of mixed farming was developing with dairying beginning to emerge as an important phase of the agricultural industry.

New and improved equipment was coming into general use; the list of such equipment now including the iron plow, potato plow, drill plow, side hill plow, harrow, drill harrow, cultivator, wooden roller, revolving hay rake, reaper, thresher, fanning mill, straw cutter, and smut machine. The agricultural societies were well established and a periodical, the *Canadian Agriculturist*, had begun publication in 1849.

The relative importance of the various crops and classes of livestock is indicated by the data in Table I. For comparative purposes data for 1966 are included in the table to show the shift in emphasis on certain crops and classes of stock since the mid-1800s. There has been a significant growth in most phases of production. The main exceptions are in the horse and sheep populations and in the production of buckwheat, each of which has shown a decline. Wheat production has stabilized. Fruit and vegetable production, not shown in the table, have become important items of commercial production and the province has developed the most diversified agriculture in Canada.

Table I
Population and agricultural statistics for Upper Canada and Ontario

		1784	1824	1826	1834	1842	1851	1966
Population	No.	10,000	150,000	166,379	321,145	487,145	952,004	6,960,870
Cultivated Land	Acres	--	--	599,744	1,004,779	1,751,528	3,705,523	12,004,305
Horses	No.	--	--	23,866	43,217	113,647	201,670	75,355
Cattle	"	--	--	114,169	179,073	504,963	744,264	3,136,956
Sheep	"	--	--	--	--	575,730	967,168	265,396
Swine	"	--	--	--	--	394,366	571,496	1,935,595
Wheat	Bus.	--	--	--	--	3,221,989	12,682,550	13,723,000
Barley	"	--	--	--	--	1,031,334	625,452	6,888,000
Rye	"	--	--	--	--	292,969	472,429	1,275,000
Oats	"	--	--	--	--	4,788,167	11,395,467	89,744,000
Buckwheat	"	--	--	--	--	352,789	679,635	14,000
Corn	"	--	--	--	--	691,359	1,688,805	59,348,000
Potatoes	"	--	--	--	--	8,080,402	4,973,285	10,584,000
Tobacco	Lbs.	--	--	--	--	--	777,426	--
Hay	Tons	--	--	--	--	--	693,727	5,456,000
Butter	Lbs	--	--	--	--	--	16,061,532	110,689,000

Data for the years up to and including 1851 are from *Census of Canada* 1665-1861. Taylor, I.B., Ottawa 1867, Vol. 4.
Data under the heading 1966 are from the 1966 census for the first six items. The remainder are 1965 data from the 1967 *Canada Year Book*.

Chapter VII
Manitoba

Early agricultural pioneering in the prairie provinces developed in different circumstances than in eastern Canada. The main influx of population into the eastern provinces began with the coming of the Loyalists in 1783-84 while the western stream commenced about 90 years later, around 1870, and swelled greatly after the construction of railways in the late 1870s and early 1880s. In the transition period between the western fur trade and agriculture as an industry, the Red River Settlement served as a testing ground, indicating the real potential for agriculture, but also showing clearly some of the hazards to be faced and overcome if agriculture was to be successful.

One big difference between pioneering east and west was the technological development that had taken place, providing the western pioneer farmers with equipment far superior to anything that had been available to the eastern pioneers. A second major difference was that the western pioneers were not faced with the heavy forest cover that had to be cleared before farming could be undertaken. In fact, the early prairie settlers were faced with a shortage of wood for lumber, fuel, and fencing. This was one of the reasons, though not the only one, that the very first settlements were strung along the rivers where there was some tree cover. Settlement on the open prairie was a second step.

Counterbalancing the advantages was the disadvantage of distance to market and, in the early years, lack of efficient transportation within and out of the area. Rivers and lakes provided means for transportation within the area, but not for export to eastern Canada or Europe. However, this latter factor did not prevent substantial immigration into the area and growth of agricultural production quite some time before the arrival of any railroad.

The exploration of western Canada was the work of the fur traders and they also were the first agricultural pioneers, although this role traditionally has been given to the Selkirk settlers who came to the Red River in 1812. By the time that they arrived the groundwork had already been done by the fur traders. In fact, the decision to send settlers was based in part on the information already available about the agricultural potential of the area. While the Hudson's Bay Company had instituted agricultural activity in the James Bay area at its first establishment the beginning of agricultural production in what is now Manitoba was at York Fort after its establishment in 1684.

The Trail Blazers of Canadian Agriculture

The early records of this post seem to be missing, but it may be assumed, based on what had occurred at the earlier posts, that gardening operations would have been undertaken almost at once. It is clear that some such activity did occur within the first decade of the post's existence as well as a certain amount of experimentation. For example, in 1693, the Governing Committee sent the following instructions to the governor of the post:

"We have sent you all manner of Seeds for a garden which we Doubt not but you will Improve now you see the conveniency of Rootes and Gardening herbs with a book giving the best directions how to use them. You must sow them in sundery sorts of Ground that you may see which will prove best to bring them to perfection; And to raise a heighth of Ground or Hedge of reeds or some Fence to keep the Norwest wind from them and then there is noe Doubt of theire comming to perfection as well as in any pte of Sweeden & Norway you have allready had great benefitt by what you have sewed."

"And likewise we recommend to you sowing Corne we have sent you of all graines some as for wheat you may sow a Gallon in one place and a Gallon in another. As soone as the ships come it must be sowed under furrow as Termed in England that it may be att least 4 or 5 inches under the earth when dug and laid hollow Lett the event be what it will matters not. You will see in Aprill whether it comes up or not if it doth come up and proves to rank when you see it soe you may mow the top of it of; if it be not rank then lett it grow in God's name."

"Then in the Spring as soone as you can gett the spade into the ground sow againe of wheat Barley Oats Beanes and Pease but must not be soe deep in the Earth not above 2 Inches and 1/2 & sow them in severall places & when you see them come up you must brake the Clods and make them lye fine but the beanes and pease you need not be so curious in lett the event be what it will lett it be done."[1]

The success in these early operations is not indicated, but in 1715, under date of September 14, the post journal records: "I draw'd the greatest part of the turnips to Day & there was abt 7 Bushells of Good Turnips and I reckon there is about 6 Bushells more Standing which I design to lett stand a little longer to see whether they will grow bigger although wee have had very cold weather & great frost for the time of year." Leaving them to grow appeared to be a mistake for on September 17: "I fear I shall loose the Turnips as is left in the Ground it being froze so hard there is no getting of them out." However, he managed to salvage the situation for on the 19th: "I had the remainder of the Turnips taken out of the Ground but it was with great difficulty the Ground being froze so hard." Then on the 20th: "I draw'd the rest of the Turnips & find our Crop has been abt 14 bushells besides what wee Spent at times which might I believe (be) abt 10 Bushells."[2]

The following year seemed to be equally successful for on September

22 the journal recorded: "I had the Turnips drawn to Day out of ye Largest Garden and find with what I have Spent and what I have is abt 24 Bushells of Turnips and are pretty Large for the most part. I have 2 small Gardens to draw more but have not many their." On October 4: "I had ye coleworts gathered and hung up for ye winter."[3]

In the early records of both York and Churchill there is no record of actual grain production although seed had been provided and instructions given for seeding. In 1733, at Churchill, reference to grain appears when the journal of April records: "Employ'd our Carptrs, In making a Roler to Levell Some Land when open Weather offers, yet we Dugg Last fall which I purpose to Experience if Barley or Oats will Grow." This was followed up with action for on April 23 it is noted: "sowed about a half an acre of Ground wth oats & barley."[4] The notation would indicate that earlier experiments had not been successful and that the governor at Churchill was unaware of earlier attempts.

In preparation for further agricultural activity it appeared that new, improved equipment was to be forthcoming for on September 24 we see: "One of our carpenters is making of a Plow & Hairah, the Smith a making Ironworks for the same." A few days later it is noted that "our Smith is making a Plow Sheer." The plow was put to use the following spring for on May 13, 1734: "we plowed a bitt of Land & Sowed the Same with Barley."[5] This very likely was the first plowing done in the province of Manitoba.

The London Committee also was concerned about livestock at the York Fort for in 1693 the following instructions were sent. "If you have many Wolfes and ravenous Beasts that will Destroy Cattle about the Factory wee recommend to you to destroy them totally, as alsoe to encourage the Indians to doe the like, for we are resolved by Gods help to send some Cowes and Goates and Swine next yeare not doubting but they may be maintained and there very well & therefore desire that a Convenient house in some Dry warme place may be built to putt them in against they arive provided you doe not find upon mature deliberation the climate alltogether impossible for theire subsistence & that one or two with Doggs and guns may attend them in the Day time as long as they can feed abroad and housed every night and watched by one or two for theire better Security & that food may be provided for them as soone as they arive next Yeare. We have sent you Sythes forks etc. to cut and make hay where we understand there is enough not far from the Factory for one Hundread head of Cattel and more and therefore we expect you to make provision accordingly for 16 or 18 Cowes for the hole winter till spring comes againe & we know if we keep great Cattell Swine and Goats may easily be keept which if we can bring to pass as we are moraly ashurd for Findland and Lapland are as cold and colder, more barren & unfruitful & much longer Nights than at York fort and yett at those places Cattell are maineteined and it would be of great advantage both to the Factory and us to have Cattell mainetained at York fort in case of any miscarriage of our ships as also the benefit you will have of fresh provisions."[6]

Apparently the Governor at York Fort was not completely convinced that the proposal was sound for the next year it was noted: "We have declined sending over Cattle for the present as you desire of us."[7]

It is difficult to pinpoint the arrival of the first domestic animals in what is now Manitoba, but it can be inferred that the first arrived at York in 1714. The journal of October 18, 1714 notes: "men cutting grass for the cattle — others clearing the Garden Ground."[8] On May 9, 1715, it was noted, when the river flooded and caused problems, that: "excepting on one of the flankers as I had left the apparently Sheep & Goat on, found that the Sow bigg with Pigg a Boar & ye Goat had jumped of wth the fright & were all drowned as likewise the hens."[8]

Apparently the remaining animals were more of a nuisance than their worth and action was taken to eliminate them. "Sept. 29 I killed my Sheep & Ram not thinking it worth while to Raise any of that Sort of Cattle the winter being so long here and the Summer so full of Musketos that the poor creatures lives in Misery either being froze in Winter or Eaten full of holes in ye summer by the Flyes so that they are all Summer Madde & all winter creeping into holes. I killed two Shoats for the same reason & design to kill all the rest next week for the charge & trouble in keeping them does not Counterwail the Profitt of them & being least 6 weeks Winter more than there is in the Bay and for want of Berrys & Roots with fish here I think not worth while to keep them."[9]

The final act was noted on October 5, when:"I killed a Large Sow and 3 Small Shoats which is all my Stock for I find there is no fish to keep them in the Long Winter & they would be more Charge than profit to us."[10]

The next attempt at livestock production in Manitoba was at Churchill in 1731. Norton, writing from Churchill states: "We also bought at the Orkneys for the use of the Company one horse, one bull and a cow, all of which arrived safe at Churchill, but soon after the horse expired with dropsy in his heart as our surgeon informs us. If this creature had lived we are persuaded it would have been of infinite service to us in forwarding of the Company's Affairs. The two latter is both hearty and well and we are in hopes will continue so, we being in a capacity to provide good quality hay for them."[11]

Early Livestock Production Efforts

It would appear that this was the first horse to come to Manitoba, predating by about ten years the horses brought back by the Verendrye sons in 1741 on their return from their trip into what is now South Dakota.

Norton was convinced that cattle would do well in the area and in 1732 he wrote to headquarters as follows: "We humbly recommend to your honour's consideration the sending over some cattle which we find should live in this country by the experience we have had of the bull and cow we bought at the Orkneys which is both in heart, likewise a calf which was brought forth the 6th February. This shows, Gentlemen, that this climate will agree with English cattle, Orkney do. being so very small and not fit for labour, and we have made use of opportunities to drain and clear some ground for pasturage which produces good grass for hay."[12] This very likely records the first calf born in the province.

Apparently his recommendation was accepted by the authorities for on August 5, 1733, he noted in his journal: "this Day We gott Ye Cattle & Everything that belong'd thereto a shore from onboard ye Mary."[13] Later he noted, "Wee also are Breaking a draught of Cattle to Draw Stone," and "others are Endeavering to bring one Oxen & Bull to Drawing, but they are Very unruly as yet."[14]

In the meantime the Orkney cattle did not fare too well. On March 2, 1733, the journal records: "This morning we Kild our Bull he having been out of Order 5 Days Past, & upon Opening him we found he had Eat Something that had Eaten a hole in his Maw of which he would certainly have Dyed had we not kild him which I did to make the Best of him."[15] The cow survived a little longer but on September 22, "our Orkney Cow being old & unfitt for service wee therefore Kild her to make the Best of her."[16] Earlier in the year she had produced a calf, the journal of May 26 noting: "this Day our cow brought into ye World a maile calf."[17]

Though no reference to the arrival of horses at Churchill was noted, other than the one from Orkney, the journal of 1734 makes several references to the use of horses so these must have arrived within this same time frame. In this connection it is interesting to note some of the items included in the invoices of goods for Churchill.

"1733 — 2 pair of Harness for horses
2 yokes & cl for oxen & 35 Bows for Do.
1735 — 1 Bridle and Whip
1736 — 1 Saddle
1 Set Waggon Wheels
2 hames
Horse nails
1738 — 6 Collers for Cart horses
2 Cart Sadles & Crooper & Gril Strap
1 horse hide
4 pair hames
1 Box Horse Medicine
2 pair Cart Wheels
2 Horses
12 Bows for Oxen
2 Cart whips
1739 — 1 Farriers shoing hammer
1 Pair Sheer's Farrier
1 plow with 6 Irons."[18]

It was noted earlier that the first experiences with livestock at York were not very satisfactory and were discontinued. It has not been possible to ascertain definitely when livestock operations were renewed at that location. The earliest record is 1746 when the journal records: "Ford, and Jayner a making a house for the Hogs."[19]

It was not until 1751 that reference was made to hay making from which one may infer the presence of cattle though they were not mentioned specifically until 1752 when the October 13 journal records: "the rest cutting firewood, and fetch some hay up for the cattle." The October 20 entry "house carpenter Enlarging the cow house" would suggest that the herd had grown and that cattle had been there previous to that time. It would seem that cattle continued to be maintained at this location as reference is made to them quite regularly. For example, on September 29, 1774: "the House & Ship Carpenter Sawyers, and part of the Labouring People at work laying the foundation of a new Cowhouse 32 feet long & 17 feet wide in the clear."[20]

Raising cattle was not without its problems and in 1777 a shortage of hay necessitated some reduction in the holdings to prevent the cattle from starving during the winter. Journal entries noting action on this provide a hint that a fairly substantial herd was in existence as well as a number of swine. Thus:

"Oct. 9 — the Steward and four Men killing and dressing a Bull, weight Neat 418 lbs.
 11 — four Men killing and dressing a Cow.
 15 — four men killing and dressing a Cow, weight neat 360 lbs.
 20 — the rest killing & dressing three hogs, neat weight 224 lbs.
 22 — the rest killing & dressing 9 piggs their neat weight 330 lbs.
 28 — the Steward, Cook, Cooper & two men killing and dressing 5 hogs, their weight neat 332 lbs.
Dec. 18 — and killing 2 Sows and 6 small pigs weight neat 363 lbs."[21]

Some of the special problems that were encountered are indicated by the following two journal entries: January 12, 1778 "this morning we had the misfortune to loose a fine yearling calf which was drownded in the watering hole,"[22] and on October 19, "Early this morning the Dogs got to the Hogs and unfortunately tore four fine piggs to pieces, this is a great loss, as they would have weighed if fatt at least 200 lbs. Another loss of five hens and a cock happened to us last week they were killed and carried away by Martins, since when we have caught four of those animals which is some recompense. Killed five small roasters their weight 48 pounds."[23]

One may wonder what feed was provided for swine in view of the lack of grain production on any scale. Fish were one source and a journal entry March 30, 1778, gives further information: "sent a man & 2 dogs with them (the hunters) to bring home the heads, wings, and guts of the partridges for food for the Hogs & dogs."[24]

Returning to the subject of crops we find that the first reference to potatoes in Manitoba is found in the York journal of May 27, 1760:

"Sowed Beans, and all Garden Seeds/ie/ I brought out Last year potatoes, Sage, Mynth, Balm all growing when arrived tried the same in this ground, but none stood the winter."[25] The next reference to this crop was in 1776 when on May 20; "Sowed several sorts of Garden Seeds and planted about 20 Square yards of potatoes."[26]

It may be argued that all of these developments took place in a part of the province that not even today has any real significance from the standpoint of agriculture. This is true, but at the same time, if we look at this in relation to some of the early developments in eastern Canada, the scope of operations was on a comparable level. However, from the long term standpoint, developments in the southern part of the province were of greater significance in providing background for future development. Here the first action did not occur until the arrival of the French from eastern Canada. There is some indication that there was some penetration into this area before the arrival of La Verendrye, but his exploration marks the first authenticated and recorded entry. After having established his staging fort on the Lake of the Woods, he moved on and established Ft. Maurepas at the mouth of the Winnipeg River in 1734 and then in 1738 he built a replacement for it at the junction of the Assiniboine and Red rivers, Ft. Rouge, and also established one in the Portage la Prairie area. There are no references to gardens being established at these sites and, in view of the fact that La Verendrye made note of his agricultural activity at Ft. St. Charles on the Lake of the Woods, the absence of such notation in his later reports leads one to question such activity at any of the later posts.

In 1793, the Hudson's Bay Company established a post at Brandon. Notations in the post journal indicate that gardening became an important activity early in the life of the post. Thus, on May 5, 1795: "4 men digging in the Gardens." May 6: "4 men digging in the Gardens."[27] Nothing is noted about the success of this endeavour that year, but in 1796 the September 27 entry states: "Took up the Potatoes a fine crop indeed and which Miller says is the produce of 12 sets only, got from the Canadians last spring, they are about three bushels and the largest I ever saw in Hudson's Bay, it is a known fact that anything will grow in this Country, the Indian corn is in full perfection but cannot be kept from the Indians pilfering it.."[28] The fact that seed potatoes had been obtained from the Canadians is proof that other agricultural activity preceded the Brandon post efforts.

The variability in product and productivity that still characterizes crops in Manitoba is indicated by the journal entry of October 8, 1797: "took up the Potatoes and Indian corn, the former is of the worst quality I ever saw."[29] This is quite a contrast to the statement of the previous year. 1793 apparently was a bad year as the October 3 journal entry states: "Henry Lena taking up potatoes a very poor crop indeed."[30] Some indication of the extent of production is given in 1800: "finished Ye

Potatoes about 50 Kegs a poor crop for ye large space of ground. I suppose it is owing to ye dry season, there was no fruit of any kind this summer."[31]

Conditions improved and production expanded for on October 16, 1806 it was noted: "finished taking up the potatoes near 300 ten gallon Kegs."[32] The next year, the potatoes were: "the largest I ever saw at this place, and there is the best onions and cabbage this Country ever produced."[33]

At Brandon, as at Churchill, the need for a plow became evident and on April 21, 1808, the journal recorded: "Alexander Costie and G. Mowatte making a plow."[34] There was some weakness in the implement for the next day it was stated that the plow broke. Despite this setback it must have performed adequately after repairs for on May 5 "Yorston plowing the Garden the second time" and on May 7, "men sowed 140 gallons of Potatoes,"[35] which would indicate that a substantial area was in production.

In 1809, a scourge, that was to recur frequently in the years to come, made its appearance as noted on September 14: "it mortified me to find that the Grasshoppers had eaten all the vegetables in my Garden except a few that were remaining in the center."[36] This may have affected the potato crop for on October 10: "Men finished taking up Potatoes and the whole amount is only 22 Kegs of Red Potatoes and 3 of white which is 78 Kegs deficient of last year."[37] The next year was better with a yield of 458 kegs, 34 of white and 424 of red. It is of interest that even at this early stage of development more than one variety was in production.

The first reference to a cart was made in 1804 and to the building of a stable in 1805. Horses were in common use by this time and were maintained in considerable numbers despite their tendency to stray and thievery by the natives.

In this same general time period there were other agricultural developments at posts of both the Hudson's Bay Company and the Northwest Company in Manitoba. It may be argued that the post at Pembina was not in Manitoba, but it had a direct relation to the area. There Alexander Henry reported, in 1802, that he had planted potatoes, turnips, onions, and cabbage. His journal entry of October 17, 1803, gives some indication of the scope of his operations: "snow. I took my vegetables up — 300 large heads of Cabbage, 8 bushels of carrots, 16 bushels of onions, 10 bushels of turnips, some beets, parsnips, etc."On October 20 he recorded: "I took in potatoes — 420 bushels, the produce of 7 bushels exclusive of the quantity we have roasted since our arrival (from Lake Superior) and what the Indians have stolen, which must be at least 200 bushels more. I measured an onion, 22 inches in circumference; a carrot 18 inches long, and, at the thick end 14 inches in circumference; a turnip with its leaves weighed 25 pounds, and the leaves alone weighed 15 pounds. The common weight is 9 to 12 pounds without leaves."[38]

Henry also was responsible for the post at Portage la Prairie and of this he says: "At Portage la Prairie we have an excellent garden, well stocked with potatoes, carrots, corn, onions, parsnips, beets, turnips, etc., all in forwardness and good order. Cabbages and melons do not turn out so well as at Pembina river the soil is too dry and sandy."[39]

In this same time period there were additional developments in what is now the agricultural area of the province. For example, Harmon, in his journal of October 10, 1800, states: "In the morning we crossed Swan Lake, which is nearly eight Miles long and then entered what is called Great Swan River (to distinguish it from the other of the same name) and is about eleven Rods wide, & plenty of water, an(d) no Rapids to this Fort which stands about twelve Miles from its entrance. The Country lies flat and low & I am told that the soil is excellent. Here we find Monsr. Perigne in charge of the Fort, where they have a tolerable Kitchen Garden."[40]

In the south some livestock production had been undertaken, for the Northwest Company Fort La Souris had cattle before the Red River settlers arrived. How these cattle were brought to this location is not known, but it is assumed that the original animals were brought by canoe as calves from the Lake Superior area. They must have arrived some years prior to 1813 if they were brought out as calves for in that year Peter Fidler bought a bull, a cow, and a heifer for the Red River settlement.[41] One may assume that the heifer had been born in the west.

But again we turn to the north, this time to Oxford House. While gardening was undertaken there as early as 1800 our primary interest is in the arrival of livestock which was recorded on June 26, 1811, as follows: "Friday at 6 a.m. two boats arrived from the Factory loaded with trading Goods and two young Calves, a Male & Female."[42] These animals were not destined for a long tenure at Oxford House for the next year Macdonell, leading the Red River Settlers, saw the animals at Oxford House and requisitioned them for the settlement.

There is more than a hint that other livestock may have been in the southern area before the arrival of the settlers. Gabriel Franchere, in reporting on his journey in 1814, wrote: "The 30th brought us to Winnipeg river which we began to ascend and at about noon reached Fort Bas de la Riviere. This trading post had more the air of a large and well-cultivated farm, than of a fur trader's factory. A neat and elegant mansion, built on a slight eminence, and surrounded with barns, stables, storehouses, etc., and by fields of barley, peas, oats, and potatoes, reminded us of the civilized countries we had left so long ago."[43]

Not only the posts were active in agricultural production. Some of the freemen had settled on plots of land and undertaken cultivation of the land to supplement their other sources of food and revenue. A specific example of this is Jean Baptiste Roi, who gave evidence in the trial regarding the destruction of the Red River settlement. He stated: "For twelve years past I have cultivated a piece of ground of my own. Before that I was in the service of the North-West Company... I used to sell the produce of my ground to the gentlemen of the North-West Company, or the Hudson's Bay Company."[44]

All of the foregoing points to the fact that primitive agriculture, but agriculture that provided some information about the potential of the area, as well as the limitations and hazards to be faced, had been under way for nearly 150 years before the Red River settlers arrived. Sketchy as the information might be, it had demonstrated that crops could be grown successfully and that domestic animals could survive the climate. In this respect the settlers did not come into a vacuum of knowledge about local conditions, such as that which the original French settlers of eastern Canada had faced when they first arrived. Nevertheless, the new settlers were faced with major problems, including the enmity of the fur traders and the Metis, who saw in them a threat to their business and their way of life.

The story of the Red River Settlement has been told in considerable detail in readily available sources and thus will not be recounted here except in broad outline dealing only with the agricultural aspects. Much has been written about the hardships that the settlers had to endure, and there is no denying that endure they did. But it is only fair to say that, by and large, such hardships could have been alleviated to a considerable extent if they had been properly equipped. The settlers came ill prepared in terms of equipment and materials, and, furthermore, their arrival was poorly timed in relation to the time of year when land preparation, and seeding and planting, should take place. So it is not to be wondered that their endeavours did not meet with great success the first few years. The records show that seeding was late in 1812, the first year of settlement, for Macdonell reported: "My people had not wrought well — a quantity of hay was made — but they had not sufficient ground cleared for the winter wheat — so it was 7th Oct. before it could be sown, & then put under with the hoe, there being no Blacksmith to make teeth for the harrow."[45] The result was that in 1813 Macdonell could only report: "Our crops from bad Culture & the seed being old, do not promise great returns, the winter wheat being late sown has totally failed; as also the summer wheat, Pease & English Barley, of all these must be fresh seed sent us. The appearance of the Potatoes promises good returns. The Indian Corn has almost totally failed from the great drowth after planting, grubs, etc. The sowing was chiefly done with the hoe, as well as the planting — only one imperfect plough was got agoing late in the Season, there being no man here capable of making a good one."[46]

Actually the situation turned out to be better than the above would indicate for in February 1814 Macdonell reported: "The crops were not well attended — I had 5 or 600 Kegs of potatoes — some of the people who cultivated for themselves had returns of at least 50 fold for potatoes. Our turnips were of an extraordinary growth, the grain did not turn out so well from the ground being badly tilled, it was choked up by weeds which the uncommon strength & richness of the soil throws out in great force. Of the different kinds of grain sown a portion has been collected for this years seed."[27]

Operations were still at a very low level, but in 1814 seven kegs of barley, four of wheat, five or six of oats and a quantity of corn and buckwheat were seeded along with 300 kegs of potatoes. At midsummer these crops were reported to be most promising.

Strife with the fur traders hampered operations and caused some crop losses, but the settlers persevered. In 1818 they had their first almost complete failure when the grasshoppers arrived in late summer and practically devastated their crops. They managed to harvest some grain for seed for the 1819 crop but the grasshoppers, this time from the local hatch, cleaned them out. In order to have seed for 1820 it was necessary for the settlers to send a delegation to Prairie du Chien, in Wisconsin, to purchase seed. Two hundred and fifty bushels of wheat were obtained and transported, at great difficulty, to the settlement in time for seeding.[48]

Scholfield summarized the situation for the next few years as follows: "The people soon recognized the wonderful productiveness of the soil. They found that wheat sown on land which had been previously cropped would return from twenty to thirty fold, and even when sown on newly broken land would yield seven fold, and that barley, when sown on well-tilled land, yielded even more abundantly than wheat. This encouraged them to sow larger areas each successive spring. The lack of plows had greatly retarded cultivation in the early years of the colony's history, but several good crops encouraged the colonists to make plows for themselves during 1823 and 1824. The iron used had to be brought down from York Factory and cost over a shilling a pound by the time it reached Fort Douglas, and the blacksmith charged £4 for ironing a plow; nevertheless a number of new plows were ready for use in the spring of 1825, and nearly twice as much land was seeded that season as in any previous spring. The crop grew luxuriantly, matured well, and was safely harvested, although a plague of mice threatened it in the fall."[49]

Things were looking up but then disaster struck when the Red River flooded its banks in the spring of 1826 and caused major destruction of buildings, fences, and livestock. This was a serious setback and caused a number of settlers to leave the colony, but the core of the Scots remained and continued to develop the area.

Livestock were introduced into the colony from the beginning but for a number of years not much success was achieved with them. Reference has already been made to the bull and heifer picked up by the first immigrants at Oxford House and to the purchase of other cattle from the Northwest Company. Correspondence indicates that an attempt was made to arrange for the purchase and delivery of cattle from Prairie du Chien to consist of "one hundred young milk cows & 4 or 5 bulls, but not to exceed one thousand pounds for the whole expense including purchase, & the expense of driving."[50] Apparently nothing came of this and it was not until 1820 that a small addition of about 19 cattle arrived from the south. In 1822 a large herd of 300 cattle was brought in and sold at £18 for good oxen and £30 for cows. Two years later another large herd of cattle was driven from the south to the settlement.[51] These herds provided the basic stock for the future herds of the colony and, later, for the province.

The first sheep for the colony were sent out in 1812. Selkirk wrote: "You will receive by the K. George a small parcel of Spanish sheep, which as well as a similar parcel in the R. Taylor I am anxious to send up to the settlement without delay. They are of a most valuable breed, selected with great care, & if lost it will not be easy to replace them another year."[52] Seventeen ewes and four rams arrived at Pembina, though in bad condition, and success was not assured as "one Ewe & 2 Rams died in the course of the winter, & all the lambs except one which is still alive."[53] By 1814 the flock was reduced to "2 Rams, 9 Ewes & 4 lambs." A major effort at enlarging the sheep population was made in 1832 when a delegation was sent to Kentucky to purchase a flock and bring it to the Red River. A flock of 1,370 sheep was purchased and in May 1833 the trek to the settlement began but the difficulties en route, especially trailing the sheep through spear grass country, resulted in the loss of practically the whole flock before the destination was reached.[54]

Apparently there were no pigs in the colony in the first few years for in 1814 Macdonell was suggesting that hogs were available at Prairie du Chien and were much wanted at Red River.

The composition of the population and the plan of settlement require some consideration in the context of the development in the colony. It has already been indicated that, even before the arrival of the Red River settlers, there had been some freemen settlers on the river, and more continued to settle there. In addition others settled at Pembina, which later was found to be in United States territory, and at White Horse Plains on the Assiniboine. All settlement occurred along the rivers, the primary transportation routes, and when the Selkirk Settlement was started, the land was laid out in narrow lots, reminiscent of the French settlements on the St. Lawrence. Basically these were 100 acre lots but behind each lot was a 'hay privilege' two miles in length to which the owner of the river lot had sole right.

Most of the freemen were of mixed blood, French half-breeds or Metis, and English-speaking half-breeds. To these were added some of the retired factors and servants of the Company, especially after the merger of the Hudson's Bay and Northwest Companies. Up to about 1830 there was then a conglomerate population of mixed-bloods, Scottish and Irish immigrants, some French Canadians who arrived in 1818, and some Swiss and Germans. Most of the latter group left after the floods of 1826 and after that about the only additions from outside the territory were 13 families from Lincolnshire in 1836 and Chelsea pensioners, old or disabled soldiers, in 1840 and 1850. Of this population only a small proportion, primarily the Scots and the French Canadians, were farmers in the true sense of the word. Most of the mixed bloods continued as buffalo hunters, or as boatmen for the Company, and later as carters.

In 1821 the population was estimated to be "221 Scottish settlers, 65 de Meurons, 133 Canadians, a total of 419 people, of whom 154 were females."[55] There are conflicting reports for 1822, but the census, as cited by Buller, showed a population of 581, consisting of 234 men, 161 women, and 286 children. There were: 3 bulls, 45 cows, 39 calves, 6 oxen; 10 sheep and 1 ram; 12 pigs; and 78 horses. Seed sown in the spring included: wheat 236 bu.; barley 15; Indian corn 13; potatoes 570, and peas 18.[56] (These data have all been rounded from the original.) Development up to about midcentury is shown in Table I.

Table I
Summary census statistics for the Red River Settlement[57]

	1831	1834	1838	1843	1846
Population	2,417	3,356	3,966	5,143	4,459
Horses	410	630	1,133	1,570	2,316
Oxen	887	1,592	1,707	2,307	2,240
Other cattle	2,066	3,411	3,748	3,894	3,654
Pigs	2,362	2,053	1,698	1,976	3,738
Sheep	--	--	457	3,569	4,188
Plows	187	275	382	429	438
Harrows	243	353	476	536	541
Acres cultivated	2,152	3,230	3,862	5,003	5,199

The data in Table I show that there was gradual growth. The major problems centred around the isolation of the settlement from the rest of Canada and even from the settlements in the United States. The supply routes continued to be through the Hudson's Bay and the canoe route from Montreal which limited the amount of equipment and supplies that could be brought in and effectively eliminated the export of any possible agricultural surplus.

The original purpose of the settlement had been to develop an agricultural supply base for the western trade and in this it was partially successful. The Hudson's Bay Company did in fact make substantial purchases. An example is the order placed in 1831:[58]

The Trail Blazers of Canadian Agriculture

```
500 cwt Flour                     10/ per cwt
200 Bushels rough Barley          2/ Bushel
 40  "     hulled Indian corn     4/  "   "
 30  "            Pease            3/  "   "
  7 cwt prime Beef                3d. per lb.
 50  "   Pork                     2d.  "   "
600 lbs Ham                       4d.  "   "
 30 Firkins Butter                7d.  "   "
```

An order, in 1845, included, in addition to items as above, "220 lbs. cheese, 15 Kegs eggs, 8 Bus. Onions."[59]

Unfortunately there were problems with the quality of the products, particularly the flour and butter, and the factors at the posts were not satisfied. Part of the problem derived from the production practices, especially the threshing procedures, which resulted in high moisture grain, which when made into flour, did not keep satisfactorily. The Company tried to overcome the problem by buying the grain and doing its own milling, but this did not improve the situation.

From the standpoint of the producers another problem was that there was no alternative outlet for their products, and thus they had no real bargaining power with regard to price.

From the very beginning of the settlement there had been tenuous contact with points in the United States and as settlement in the States moved northward this contact increased. Thus, beginning in 1844, there developed a regular trade with St. Paul with cart trains being the form of transportation. "Six carts were reported in that year, 102 in 1851, 600 in 1853, and by 1869 the number reported was 2,500."[60] Competition with the ox cart began to develop in 1859 when the first steamboat was put into service on the Red River and from that time forward river traffic increased and became the main form of transportation until the railway connection was made to the south in 1878.[61] Shortly after, direct railway connection to the east was made and full contact with eastern Canada was available.

An interesting aspect of agricultural development was the establishment of experimental farms by the Hudson's Bay Company. The first one was established in 1821, but it was discontinued shortly after without having contributed anything to development. The next one was begun in 1831 at which time new farm equipment, implements, and brood mares were brought in. This one failed and then a third one was started in 1838. According to Ross: "The only benefit settlers derived from the example of the experimental farmers, and what they had not learned themselves before, was to mow down their fields of grain with the scythe, in place of cutting it with the sickle; and to gather it with rakes in lieu of tying it into sheaves."[62] In any event this farm did not prove to be any more successful than the other two had been.

Government Regulations

While agriculture to mid-century basically was subsistence agriculture, efforts were made to improve production methods and the quality of the livestock. The Council of Assiniboia, the governing body for a number of years, took action to regulate and stimulate the industry. For example, at a meeting on May 4, 1832, the following resolution was passed: "That all persons be at liberty to seize any pigs they may find trespassing on their lands, whether those lands be fenced or unfenced, to retain possession of such pigs until the parties to whom they belong pay the sum of two shillings to the proprietor of the ground on which such pigs are found for every trespass, and unless that sum be paid within eight days after such pigs may have been seized, the parties be at liberty to sell them after giving eight further days notice to six constables that such pigs are in pound; and that all constables be hereby authorized to seize any un-ringed pigs they may find straying after the 10th of May, beyond the boundaries of the property occupied by the parties to whom such pigs belong and to retain them for their own benefit, as a perquisite of office."[63]

At the same meeting there also was a resolution controlling stallions running at large and a further one that stated: "That public fairs shall be held annually at Frog Plain on the 1st Monday after the 20th September, and on the 1st Monday after the 20th of May ensuing."[64] An earlier regulation set the time at which the cutting of hay on the open prairie could commence. The purpose of this was to give everyone an equal opportunity. The dates for commencement of cutting were set annually.

In 1839 the Council resolved: "That a premium of five shillings be paid for the head of every wolf killed within five miles of the banks of the Red River settlement, the head to be delivered to the Chairman of Public Works on receipt of the premium."[65]

From time to time new regulations were made regarding animals running at large, but a more positive move for improvement is seen in action taken in 1845: "Whereas this secluded Settlement must, in a great measure, rely on its own internal resources, as distinguished from foreign trade, it is Resolved, 8th, That the Bishop of Juliopolis, the Rev. John Macallum, Captain Cary, Dr. Bunn, and Mr. Pritchard, be a Committee of Economy with power to encourage by premiums or otherwise the improvement of such branches of agriculture as may bear on such improvements, either by producing materials or by saving time, and it is Resolved, 9th, That the said Committee may import, duty free, such seeds and drugs and implements and other things as may appear likely to be beneficial, and shall sell the same at cost and charges, under proper guarantees against their being resold."[66]

Arising from this action, the Committee, in 1847, "awarded prizes including £5 to the person that produces the largest quantity of the best quality cheese."[47] In the same year, owing to the deterioration of the grain in the settlement, the Committee ordered the importation of 100 bushels of Black Sea Wheat."[67] In 1850 the Council agreed to provide a grant of £25 for the newly formed Agricultural Association.[68] This Association had as one of its objectives "to encourage stall-feeding and other branches in which the farmer is deeply interested."[69]

Other action also had been taken. Without being specific, Ross stated that "costly bulls, and of the finest breeds, both from England and the United States, have been imported into the Colony."[70] Campbell, in 1833-34, made specific reference to the English Hackney stallion "Fireaway" which had been imported to head up the Company's band of horses.[71] There is no statement as to how it was brought in. Ross, about 1855, stated "another full-blooded stallion, however, came out from England, a year or two ago, at a cost of £300, superior in size to his predecessor but inferior in model and action."[72]

Despite such action the cattle apparently deteriorated in size. One of the reasons given was lack of care and the second was that "the local government has taken no steps to restrain a multitude of dwarfy bulls from running at large in all seasons."[73] Sheep were said to be declining fast in numbers from the ravages of dogs and wolves.

For various reasons the decade 1850-60, was a period of subtle unrest with a portent for the splurge of immigration and development that commenced in the 60s and came into full force in the 1870s. First there was the influx of traders, vying with the Hudson's Bay Company and leading to increasing trade, by way of the ox-cart trains, with St. Paul. There was an underlying dissatisfaction among many of the young people with the lack of opportunities in the colony. Then there was the increasing interest of the Imperial government and the governments of the Canadas in the vast territory of the west, which led to the exploring parties of Dawson and Hind, and that of Palliser, all of which provided enlarged knowledge of the natural resources of the area.

Of direct influence on agriculture was the importation of new equipment including such relatively advanced machines as the reapers. This enabled farmers to increase their acreages and their productivity. Until that time the relatively primitive and poorly developed agriculture had been able, coupled with the buffalo hunt, to provide sufficient food for the colony and the fur trade. However, it had not been able to meet the increased demand created by the military forces sent there in 1845 and the later exploring parties.

A second major flood struck the colony in 1852 and again caused widespread destruction and hardship. This led to the first major hiving off from the colony and the first westward thrust of settlers. In 1853 Reverend William Cochran led a migration of settlers from St. Andrew's and Middlechurch to the higher ground at Portage la Prairie. This move had later important political repercussions, but agriculturally it was a factor in compounding land claim problems at a later date. Actually the only people who had any semblance of legal rights to land were those settled on the Red River who had conditional long-term grants or entries in the books of the Hudson's Bay Company. All others were squatters who held the land merely by occupying it.

The filling up of the better lands of Ontario was adding ferment as younger generations of Ontarians began seeking new lands and followed the exploring parties to the Canadian prairies. Some of these people settled in the older settlements on the Red, but most of them went on to the newer settlement at Portage where land was more readily available. All of these settlers followed the pattern of locating along the rivers in order to have access to water transportation and wood. In other words, the treeless prairie remained unsettled.

In the 1860s the trickle of immigrants continued despite years of drought, hail, mildew, grubs, and grasshoppers. In fact 1868 turned out to be a disaster year with a complete crop failure and it was only through financial aid from the Council and charitable donations from eastern Canada and the United States that the settlers survived. One of the early arrivals was a John McLean from Canada, in 1862. He purchased land from previous settlers, paying them only for their improvements. This was the generally accepted practice for those who occupied land by squatter's rights only, with no legal title to the land. Kenneth McKenzie, from Ontario, made an exploratory visit in 1868, and, despite what he saw in that disaster year, decided that there was great potential. The following year he arrived and passed through the settlement at Portage and "ploughed a challenging furrow around fourteen hundred acres of grass and timber land at Rat Creek."[74] He is reported to have brought the first Shorthorn bull to the Northwest.

In the 1860 decade the territory of western Canada was acquired by the Canadian government from the Hudson's Bay Company. It was recognized that an official survey of the land would be necessary in order to provide legal title to the increasing number of immigrants; therefore, a survey was begun in 1869. However, this was interrupted by the political upheavals that took place, namely the first Riel rebellion. It was not until 1871 that the surveys were renewed, but then they proceeded with despatch so that by 1873 all of the province, as then defined, was surveyed into the familiar square township pattern with special provision being made for the river lots in existence. In the interim the new settlers had been assured that if they registered their claims these would be recognized for title after the survey was completed. As a consequence, the newspaper columns were filled with notices of claims. There had been disputes over claims such as where the Pembina trail crossed the stream now called the Boyne but first known as Riviere aux Ilet de Bois. A group of Metis had settled here, but apparently left during the summer for the buffalo hunt. Settlers from Ontario were attracted to this spot by the presence of water and wood, as well as the soil type, and, ignoring the Metis occupation, staked claims. No open conflict ensued as the Metis gave way.

The railroad facilitated settlement of the west by moving people and agricultural products could be moved to market. However, there was a relatively large influx of settlers into Manitoba, and some farther west, before the coming of the railroad. They came by all other means of transportation. A large proportion came from the east by rail through the United States to the point of river transport north, but substantial numbers came by the Canadian route of steamship to Fort William and by road from there or by ship to Duluth and then by train and river boat to Winnipeg. For example, the *Manitoban* reported, on April 19, 1871: "It is said that 2,500 tickets have been sold in Toronto to emigrants coming here by the Canadian route."[75] Throughout that year and the following years there were frequent newspaper accounts of the arrival of new immigrants and their belongings by steamboat, barge, and land transportation.

Disposal of surveyed crown land, beginning in 1872, fell into three categories. The first was the homestead of 160 acres at a cost of $10 registration and the obligation to live on and develop the land for five years, later reduced to three years. The second method was the pre-emption of an additional 160 acres at $10, and finally there was cash purchase. No more than 640 acres could be allotted to any one person. People could also purchase from original settlers or from the Metis for whom 1,700,000 acres were set aside. Additionally, the Hudson's Bay Company had retained some land and the railroad was given grants of land which were all available for purchase.

The land office started issuing homestead and pre-emption rights in 1872 with John Sandersen getting the first homestead on the Portage plains.[76] Business apparently was brisk for the newspaper reported, on July 6: "a large number of business is being transacted at the Dominion Land Office, in entering homestead and pre-emption rights in townships already surveyed. So great is the rush, that the officers have to work early and late to keep up with their business."[77] Action continued brisk for several years and on May 2, 1874 the paper reported: "The business of the Land Office during the past month has been more than double that of the previous month of this year, or the same month of last year."[78] Influx of settlers continued for Begg wrote: "in 1878 the population of the country had doubled from what it had been in 1871 and settlement was rapidly spreading westward beyond the boundary of Manitoba (i.e. the original boundary). At the close of 1879, farm houses and cultivated fields were in sight all along the main road 250 miles west of Winnipeg."[79]

Settlement Encouraged

This report is corroborated by the statement in the *Saskatchewan Herald* in 1878: "It seems but a little while ago when the Portage (la Prairie) was considered to be the last point to which civilization extended, and at which home comforts could be had. Now, however, a man may leave Winnipeg and find a house to sleep in every night until he passes Shoal Lake — a distance of two hundred miles and this year the line of settlement will extend still farther westward."[80]

The first major move away from the river settlements had come in 1874 when a colonization party from Michigan and Wisconsin took up two townships on the United States border, east of the Red River, to be known as the Emerson settlement. This was the year when major group settlements began — settlements of people neither English nor French. The first such group was the Mennonites who were settled on a reserve east of the Red River. In 1876 another reserve was set aside for them on the west side of the river, taking in the area where the towns of Altona and Winkler now are located.[81] The first group of Icelanders came in 1875 and settled in an area which at that time was outside of the then established boundaries of Manitoba.

Efforts were made to maintain a substantial French presence in the province by enticing French Canadians from Quebec and the eastern United States to move to the province. This effort was only partially successful, especially in getting immigrants from Quebec, but about 2,000 came from the United States between 1874 and 1887.[82]

That widespread interest was being generated about Manitoba in the east during the early 1870s is indicated by a newspaper item that appeared in 1875: "Manitoba is an agricultural country. According to the *New York Independent*, Manitoba which has hitherto been esteemed chiefly on account of its fur business, has fine agricultural capacities. It is said to be one of the finest wheat growing countries in the world, the soil being black, alluvial mold, rich in organic deposits, and resting for a depth of from two to four feet on a tenacious clay soil. Some of the fields on the Red River produced forty successive crops of wheat without fallow or manure. The yield has reached as high as fifty or sixty bushels, even under the farming of the natives. Last year oats averaged sixty bushels to the acre, weighing forty pounds to the bushel. One farmer raised 350 bushels of potatoes from seven bushels of seed, and the root crops were generally good. Among vegetable products successfully cultivated are mentioned turnips, parsnips, carrots, cabbages, melons, beets, pumpkins, squashes, celery, lettuce, spinach, cauliflower, cucumbers."[83]

Parenthetically, it may be mentioned that the deep, black, alluvial soil was an old lake bed, quite low in relation to the river level in the settlement area. While not evident today, because of large drainage schemes, much of this area was swampy and unfit for cultivation in the early years of settlement. Southesk, who visited the area in 1859-60, provides a picture of this as follows: "In an agricultural conversation today, it was stated that cattle have much trouble in getting food in summer, the ground being so broken up with swamps as to cause them many miles of daily wandering in search of pasturage. Inquiring if the drainage of these marshes would be possible, I was told that in itself it was the easiest thing in the world, my informant having dried a large tract by a single open drain 900 yards long, expecting also to make an extensive improvement by carrying this work two miles farther. The drain was not more than two feet wide, by a foot and a half in depth, which is sufficient in this deep, soft soil, when the floods of a single springtime will enlarge a small trench into a wide and deep watercourse; in proof of which it was mentioned that a cow had been drowned in the drain just referred to before it had been a year open. The unwillingness of people to labour at such work is the great obstacle to carrying out drainage operations."[84]

Table I shows the population of Red River to be 4,459 in 1846. In 1870 it had grown to 11,963 consisting of 1,565 whites, 5,757 Metis, 4,083 English-speaking mixed bloods, and 558 Indians.[85] After that the increase was relatively rapid. The number of immigrants was 11,970 in 1875 and between 3,000 and 4,000 in 1876. The railway connection to the south and east became available in 1879 and immigration increased to 11,500 in 1879, 18,000 in 1880, and 28,600 in 1881. By that year "the Saskatchewan trail had carried settlers to the western limits of the extended province,"[86] and in a sense the agricultural area of Manitoba had been occupied though there still remained large areas of land available.

It may be said that, by this time, the pioneering phase of agriculture was over in Manitoba though considerable time was to elapse before agriculture had reached a stage of maturity. It is opportune to review some of the developments up to this time.

One of the factors that made possible the rapid agricultural development was the availability of improved equipment to replace the homemade and cast iron plows, the sickle, the cradle, and the flail. Despite the problems of transportation, new equipment was brought into the province and as early as 1850 there were eight threshers, two reapers, and six winnowers in the valley.[87] Two newspaper advertisements, one in 1860 and the other in 1870, indicate the availability of implements: "Agricultural warehouse and seed store : Agency of the Albany Agricultural Works, St. Paul, Minnesota. D.C. Jones, Proprietor. Keeps constantly on hand, and for sale at low prices, for cash: One-horsepower thresher and separator; two-horsepower thresher and cleaner; two-horsepower thresher and separator; four-horsepower thresher and cleaner; six-horsepower thresher and cleaner; fanning mills of all kinds; one-horse mowers; two-horse mowers; two-horse mowers and reapers combined; steel plows of all sizes; harrows and cultivators; seed sowers (broadcast); seed drills; hay and straw cutters and corn shellers; grain cradles; grass scythes and scythe sneaths; shovels, spades, rakes, hoes, mattocks, and everything usually found in an agricultural warehouse; Coleman farm mill, two-horse power for grinding flour."[88] "Agricultural Warehouse of Severcool & Grovenor, St. Paul and St. Cloud. Wholesale dealers in Case's threshers, New York reapers, Hubbard mowers, Fish Brother's wagon, Esterly seeder, Buckeye grain drill, Meadow King mower, Hollingsworth sulky, hay rake and all kinds of agricultural implements."[89]

In 1871 a local Winnipeg merchant advertised: "Flower and vegetable seeds, reapers, mowers, fanning mills, straw cutters, ploughs, harrows, superior wooden pumps, churns, vibrator thrashing machines, cultivators, sulky rakes." The first steam thresher was imported in 1874.[90]

The seedsmen were in the advertising game as well, one interesting example being: "Everyone wants Ramsdell Norway oats. Each order is entered on our books as received, and the seed forwarded as fast as we receive your name. We wish to be prompt, but sometimes the large number of orders renders a few days delay unavoidable. When you can raise 100 bushels of oats to the acre, you are losing money to sow a kind that will not give over 35 bushels. It is as plain as that two and two make four... Price list; Per bushel $7.00; per half bushel $4.00; per peck $2.50." The address for this was New York and Chicago.[91]

Though the fur traders and the Red River settlers had provided basic information on the potential for agricultural production the settlers coming from eastern Canada had a good deal to learn after they arrived. The climatic and soil conditions were quite different than those to which they were accustomed and adjustments had to be made, both in terms of equipment as well as the general practices that were most suitable. The plow was the basic implement, but the type satisfactory in Ontario was not the most suitable in Manitoba. This is evident from a statement by Kenneth McKenzie, an early settler at Portage la Prairie, in evidence given to the Select Committee on Immigration and Colonization, in 1876. The statement, prepared in 1873, and originally in question and answer form, is presented here in summary, as it outlines quite clearly the production picture of the time.

"When the summer is wet or moist I would rather break prairie sod than old spear grass sod as we do not require to break it so deep. June and July are the months for breaking, but earlier will do if you have time, as later does not answer so well. The plow I use is American, made by John Deere, Moline, but other Americans make good breaking-ploughs — light with gauge wheel in front, revolving coulter, mould boards and coulter and shear, all steel. No use for any other material here in ploughs but steel. The soil is rich and very adhesive, and even to steel it will stick a little in wet weather, more so after it is broken and cultivated. The American ploughs answer for both ploughing and breaking at present. I have a Canadian plough which does very well, but I think a good light Canadian, all steel, or even glass mould-board, would be better after the land begins to be old or long broken. We cannot go deep enough with the American ploughs when land is getting old and needy."

"On a twelve-inch breaker, we use one pair horses, or one yoke oxen. When sixteen-inch, we use three horses or two yoke oxen. I prefer twelve-inch to larger ones. With this about one acre is a fair day's work, i.e. day after day. Some, of course, will do more. The large plough and more teams will break one and a half acres. Two ploughings for the first crop answers best, i.e. one light or 2 inch in summer, and then 2 inches more, stirred up next spring. We plough both times same way, and not cross the first breaking. I have raised potatoes and turnips last year on breaking, had a fair crop, but would not like to depend on it if the season is dry."

"This year I had, spring wheat 90 acres, barley 30 acres, oats 1 acre, peas 8 acres, rye 1 acre, flax 1/2 acre, potatoes 6 acres; the rest, roots of various kinds, and clover and timothy. Wheat varieties are Golden Drop, Glasgow or Fife, and a little Rio Grande, I think it is called. I sow 2 bushels per acre and get an average yield of 30 bushels. I have had over 40. I raised 300 bushels of onions last year, and sold them readily at $2 per bushel, and I expect fully as good a crop this year."[92]

With reference to pests he wrote: "I have heard some complaint of grubs, but have not suffered any by them on my crops. I have sown turnips in May and they did well, and all through June, and no flies to hurt. Black birds were very bad at first, especially on oats, and that is the reason I had no more sown this year."

Of equipment he noted: First class marsh harvesters, or machines which will employ two men binding and of the most improved make, are wanted. I have two combined ones, made by Sanger & Co., Hamilton, which answer well, but those that will cut wider and quicker are required. I think grain drills or broadcast sowers would be an improvement, as it is generally windy here in the spring. They should be wider than those in Ontario, say eleven or twelve feet."

McKenzie did not depend on crops only for he wrote: "I have a gang plough, but have scarcely used it yet as I have delayed breaking ground by taking in some stock from the States. Last year I imported 84 cattle, and this year over 200, from the increase of which I now have fully 300 head. I am selling some daily, and purpose disposing of all the surplus I do not intend to winter, by auction, on the day after the County Exhibition, 2nd Oct. All my cows are served with thoroughbred Durham bulls, and I have three thoroughbred calves, and one aged bull. Last season I sold at fair prices one thoroughbred bull and one calf."[93]

He was not the only one who imported cattle for even earlier, in 1871, it was reported that: "Mr. Hy. McDermot and Mr. Alex Logan, who left by the Selkirk, are on their way to the States, to purchase cattle. They will bring in 200 or 300 head. We hear also that a joint stock company has been formed for the purpose of importing cattle largely."[94]

The relatively rapid increase in population created shortages of goods and supplies in the early 1870s, until the new immigrants became established and produced crops. In 1871, the editor of the *Manitoban* commented on the high prices created by the shortages: "We must make some allowances for farmers who have been so troubled with grasshoppers as ours have been for two or three years. Then being accustomed to a limited market, they never produced as much as they could, and the habit remains in many districts, even though the market is ample. We have usually so little surplus, that the sudden addition of even a few hundreds to our population, affects prices. If thousands come among us this summer — consumers, but nonproducers as they must be — we will have something like starvation prices by next Spring, unless they bring provisions with them."[95] In light of this he suggested that prospective farmers should bring with them horses, oxen, and cows as well as a good supply of food items in order to avoid the high local prices.[96]

The *Free Press* also commented on the problem, in 1873, stating that large quantities of grain were being imported while dairy products and other necessities were in short supply even though the country was well adapted to their production. An advertisement of that time announced the arrival of 30,000 pounds of Ontario dairy butter and a small quantity of Ontario cheese.[97]

The shortage of agricultural products was a temporary affair, lasting only until the new immigrants got into production. On October 23, 1876, a most significant news item appeared in the *Manitoba Free Press*: "The first shipment of grain from this province took place on Saturday last when 857 1/6 bushels of wheat were consigned to Steele Bros., seedsmen of Toronto, by Messrs. Higgins & Young, of this city... The price paid was 80 cts per bushel, and the rate of freight to Toronto is 35 cts. The return from the various threshers of wheat sent was thirty bushels per acre average."[98] A greater amount had been wanted by the Ontario purchaser, but no more was available.

Wheat to Europe

During the next year editorials and letters to the editor discussed the economic feasibility of exporting grain, with the transportation system then available, namely, boat transport during the summer and fall, to points in the United States where transfer to rail could be made. No consensus was reached but on October 17, 1877, the paper reported the first export of Manitoba wheat to Europe: "Messrs. R. Gerrie & Co. ship today per International the first carload of wheat to the European market, consigned to Barclay & Brand, Glasgow, Scotland, the senior of the firm being a brother of Mr. James Barclay, of Stony Mountain. The quality is first-class, and suitable for either seed or milling purposes... while the quality will be found desirable for mixing with British wheat. Besides the carload for Glasgow, four carloads go to A. W. Ogilvie & Co., Goderich, the day's shipment amounting in all to 1,700 bushels."[99] A further indication of the amounts of grain being produced and transported is the report: "In June Peter McArthur's Prince Rupert had brought down from Portage la Prairie three barges loaded with 500 bushels of wheat, 3,000 bushels of oats, and over 1,500 sacks of flour."[100]

The extent of production of the various grains is shown in Table II (page 185), beginning in 1881. This shows clearly that once the farmers got into the province in any number, they soon got production underway. The increase from 1,033,673 to 18,352,929 bushels of wheat in the two decades from 1881 to 1901 must be classed as a remarkable achievement. This increase in production necessitated major changes in the system for handling grain and led to the development of the grain elevators and the companies that owned them. The first storage buildings for handling grain were flat warehouses, to be followed later by the elevators. As Strange[101] has recounted, in 1879 there was one single elevator in the whole of western Canada while in 1890 there were ninety elevators and 103 flat warehouses in Manitoba. By 1900 this had changed to 375 elevators but a reduction to 94 warehouses. Not all grain was handled through these facilities as some farmers loaded their grain directly into railway cars. They could not always get the cars they wanted, claiming that there was collusion between the railway and the elevator companies to force the movement of grain through the elevators. In any event there was great dissatisfaction and this led to representations to government with subsequent investigations that led to the enactment of regulations governing the availability of cars and other aspects of the grain trade.

Though grain production was the main interest of the farmers there was some early interest in other crops as well. For example, it was reported in 1862, that: "We understand that Mr. Rowland's gardener has this year grown a considerable quantity of excellent tobacco. This is a capital innovation. We hope it may so extend that the settlement may soon dispense with importation altogether."[102] This hope turned out to be vain, though attempts at tobacco production did occur from time to time.

Another early production activity, somewhat surprising in view of the climate, but in keeping with the interest of people from Ontario, was the growing of apple and other fruit trees. The *Free Press* reported, in 1874: "Fruit trees — In the Bishop of Rupert's Land garden at St. John's are a number of apple trees, planted a few years ago, that are doing exceedingly well. This is perhaps about the only experiment that has been made in this Province, and it has been a success."[103] That it possibly was not the only experiment is indicated by a comment in the same paper on June 6: "Small fruits — A large consignment of apple and small fruit trees was brought down the river the other day, and sold off very rapidly. A few choice grafted apple trees may yet be had on application to Mr. Laurie."[104] One might assume from this that previous importations had been made or a large consignment would not be logical. The fruit industry did not become a major factor in Manitoba agriculture but it may be noted that the statistics for 1891 recorded 713 bushels of apples produced in the province.

Earlier it was noted, in what may be considered as the colony stage, that action had been taken to form an agricultural association to encourage improved agricultural practices and that this association had received financial support from the governing council. It did not take long for the settlers, coming in the 1860s and early 1870s, to take similar action as a Provincial Agricultural Society was organized in 1871 and held its first exhibition in Winnipeg on October 4 and 5 of that year. In the report of the fair no breeds were mentioned for horses, but for cattle, prizes were offered for Durhams, Devons, and grades, though no Devons actually were shown. For sheep, Leicesters and Southdowns were shown, but no Merinos, though there was a class for them. For swine, only Berkshires and grades were listed. This indicates that breeds still were limited in number. The society held a fair again in 1872, but then discontinued at this site for many years.[105]

In 1872 the legislature passed an act for the establishment of Agricultural and Arboricultural Societies in Manitoba, with the provision for financial assistance if the societies met certain conditions, including membership subscriptions to a certain amount. In that year several agricultural societies were formed and the one that included Portage la Prairie, the Marquette Agricultural Society, staged an exhibition that fall. It was continued in 1873 and the prize list for crop products shows that numerous varieties were in use. Thus, for seeds: fall wheat; rye; spring wheat, Golden Drop, Glasgow or Fife, and other variety; barley; oats, black or white; peas, large and small; corn, large and small; flax seed; buckwheat; clover seed; timothy seed; hops; turnip seed; onion seed; carrot seed; beet seed. In the potato classes there were: Early Rose, Shamrock, Kidney, and other varieties.[106]

It was not all clear sailing for the societies. At its annual meeting in 1875, the Provincial Association reported a deficit of $166.29. The Lisgar Society was in much better shape with a balance of $360.00 but the Selkirk Society had come to grief and wound up its affairs that year, only to organize a new society to replace the old one.[107] However, the societies kept on increasing and in 1882 there were 19 local societies in the province and 17 of them held exhibitions.

In addition to staging exhibitions the societies were involved in the importation of improved livestock and seed and also in staging plowing matches. In 1874 they were called on to perform a somewhat unusual function in that, following the grasshopper plagues of 1872 and 1873, the Dominion government made a grant of $10,000 for seed wheat and called on the societies to handle the purchase and distribution of the seed.[108]

As in the east, the early newspapers, recognizing the importance of the agricultural industry, provided considerable space to articles and comments on agricultural subjects. In Manitoba most of the articles came from sources whose conditions were quite different and the information, at times, was not completely pertinent. However, some of the information was valid in the context of the knowledge available at the time. This was more true of information regarding livestock production than crop production. The following listing provides a sampling of the subjects dealt with during the early 1870s: manure for potatoes; carrots for horses; fruit trees; description of breeds of cattle: Shorthorns, Devons, Ayrshires; advantages of fall ploughing; how to fatten a horse; tree culture; making butter; how to grow sugar beets; singing to cows; to hybridize wheat; transplanting trees; feeding potatoes to horses; harvesting hay early; value of straw for fodder; breeding stock.

Grasshopper Plagues

Throughout the early history of agriculture in Manitoba, even before the Red River Settlement, the producers encountered a variety of environmental hazards, particularly to their crops. Foremost among these were the grasshopper plagues which usually lasted for more than one year, as the swarms of adults that invaded the crops deposited their eggs for a new generation the next year. Often this generation was more destructive than the original one. There were major plagues in 1819-20-21, 1857-58, 1864-65-66, and a really devastating one in 1868 when it combined with drought to bring about a complete crop failure. In the 1870s severe damage was caused in some years, but no complete crop failures were experienced. No real means of control were available to the farmers and all that they could do was to wait for nature to take its course and end a cycle. It is of interest to note that as early as 1865 an article in one of the papers mentioned parasites of grasshoppers as the cause of their destruction.[109]

The only other insect pests to which reference was made, were wire worms and grubs, but without any description of the grubs.

Among the climatic hazards encountered were frost, drought, hail, and, at times, too much rain, particularly at harvest time. Then there were the floods, mainly in the Red River valley, that resulted from too much snow in the winter and rapid thawing in the spring.

Three plant diseases were recorded, namely smut, mildew, and rust. Smut was mentioned early in the settlement period and may well have been introduced with the very first wheat as it was a common disease of wheat in other parts of the country. However, in most cases reference to it was somewhat casual, as if it was not a major problem. The same is true of mildew. Rust was mentioned from time to time but sometimes it was not clear whether the terms mildew and rust were synonymous.

Weeds of various kinds were early hindrances; in 1814 MacDonell reported that his grain "was choked up with weeds."[110] The well-known Canada thistle apparently was a major problem, sufficiently so to draw an editorial entitled, "Must settlement be abandoned", in 1864.[111] In 1871 the legislature recognized it as a problem and legislated for its control. Wild oats made a relatively early appearance and their control was discussed by 'Agricola' in 1860 under the heading "Fields overgrown with wild oats and thistles." Considering the fact that wild oats continue to be a major weed problem it is of some interest to note the control method recommended by the writer of the article: "I will now endeavour to point out the fallow which I think necessary for the destruction of wild oats. Let the land infested by it be well manured in the Fall, and trench-ploughed and left rough. Early in the spring, harrow and roll as soon as weeds and oats appear. Plough again and harrow and roll, and continue doing so, as often as the season will admit. Plough and harrow and roll the land as often as there is any appearance of the oat, and the next spring sow it, and I think wild oats will not be seen. And if some should remain, sow turnips or plant potatoes the next year and by this means reclaim perhaps the best part of a valuable farm. It is preposterous to say that a piece of land infested by the wild oat must lie lea for four years, to be destroyed."[112]

The third serious weed was wild mustard which was included in the noxious weed act in 1883.

While the early settlers in the east faced the hazard of forest fires, the prairie settlers had the prairie fires to contend with. They occurred with great regularity and caused considerable losses. For example, on April 20, 1872, the *Manitoban* commented on the hay famine, the principal cause of which was the prairie fires of the previous fall which had destroyed much of the harvested hay and led to actual starvation of cattle.[113] One might assume that the usual protective measures of fire guards had been neglected or inadequately carried out.

An unusual problem, especially in the early settlement years, and even in the 1870s, was the presence of immense flocks of blackbirds and pigeons that descended on the crops during fall migration. A graphic description was given in 1860: "The birds are very troublesome - especially blackbirds, and that pest of antiquity, the cornix improba (crow). We examined a field of barley, eight bushels sowing, and we verily believe that not one half-bushel will be returned, so completely is it destroyed by birds."[114] With increasing acreage of crops, and the destruction of the pigeons, this became a decreasing problem, though, even today, the blackbirds can be destructive in certain localities.

Another pest, peculiar to the prairies, was the gopher, which continued to be troublesome. Its importance, at that time, may be judged by an article which stated that in one case 55 acres out of 95 had been destroyed by gophers.[115] Another indication of the serious nature of this pest is that in 1889 the Legislative Assembly of the Northwest Territory asked for a bonus of one cent per head to encourage destruction of the pest.[116]

The livestock population did not escape problems though, in general, these did not seem to be so frequent or important. Mosquitoes and flies caused discomfort to the animals, especially at certain times of the year. Only four diseases were mentioned specifically, namely, 'epizootic catarrh' of horses, anthrax, mange, and glanders.

Legislation Enacted

Reference was made earlier to some of the actions taken by the governing council of the Red River Settlement to regulate and encourage agricultural activity. Later legislative bodies also took similar action and a sampling of the laws passed by these bodies gives a good indication of what problems appeared to be of importance and requiring legislative action.[117]

1871
- An act to prevent the deposit of manure on the banks of the river. This was an early pollution control law to provide for potable water from the river.
- An act concerning horses at pasture.
- An act to impose a tax on dogs in this province.
- An act relating to homesteads.
- An act concerning stray animals. This was a pound law.
- An act in reference to certain animals going at large at certain seasons. This applied to rams and stallions.
- An act for the destruction of Canada thistle.

1872
- An act for the establishment of Agricultural and Arboricultural Societies in Manitoba.
- An act for the prevention of prairie fires.

1873
- An act to encourage the planting of trees.
- An act for the protection of sheep.
- An act respecting animals running at large. Bulls and pigs were added to the original list.

1875
- An act respecting boundary and line fences.

1876
- An act respecting the Bureau of Agriculture and Statistics. This

Manitoba

actually established a Department of Agriculture.
- An act to prevent breachy animals running at large. This act defined legal fences as being at least four feet nine inches high, lower rail might be fifteen inches from the ground and no other rails more than ten inches apart.

1877
- An act to require owners of threshing and other machines to guard against accidents.
- An act regarding marks and brands of cattle.

1878
- An act to prevent the extension of prairie fires.
- An act to protect native cattle from disease. This related to imported animals.
- An act concerning drovers and traders.
- An act to encourage the destroying of wolves.

1879
- An act respecting infectious and contagious diseases of domestic animals.

1880
- An act respecting drainage.
- An act respecting line fences.
- An act respecting the herding of animals.
- An act for the establishment of Agricultural Electoral District Societies.

1881
- An act respecting veterinary surgeons. This established a Veterinary Association.
- An act for the protection of public fairs or exhibitions held by agricultural, horticultural, and industrial societies.

1882
- An act to prevent the spreading of wild mustard or Canada thistles.
- An act for the reorganization of the Department of Agriculture and Statistics. Among other things this replaced the Provincial and Industrial Society with a Board of Agriculture.
- An act to enable municipalities to perform drainage work in certain cases.

1883
- An act respecting the Department of Agriculture, Statistics, and Health. This included penalties for selling seed containing noxious weed seeds — wild oats, mustard, and Canada thistle.

1884
- An act declaring anthrax an infectious disease.
- An act to incorporate the Manitoba North-West Farmers' Cooperative and Protective Union.

1885
- An act respecting communicable diseases of animals. Scabies or

mange and anthrax were mentioned.

1886
- An act to incorporate the Manitoba Dairy Association.

1890
- An act respecting diseases of animals — glanders or farcy were mentioned in addition to those previously listed.
- An act for the protection of insectivorous birds beneficial to agriculture.
- An act respecting the establishment of Farmers' Institutes.

It is evident that, as the agricultural population increased and the industry developed, new problems arose that required legislative action. While regulatory action of specific production problems was important the development and legal recognition of organizations of farmers and professions was of equal or greater significance. These organizations were indications of a maturing industry organizing to act on problems of concern to the members, a movement that continued to grow and became of increasing importance in the next century.

By the end of the 19th century agriculture was well established in the province and the pioneering days were over. Settlement had penetrated into most of the agricultural areas of the province. Some 200 years had elapsed since the first recorded agricultural activity in the province at York Fort on Hudson's Bay. Until 1860 agricultural development had been very slow and limited but during the following 45 years there had been literally explosive growth, especially after 1880. A rail transportation system had been built to provide outlet to world markets for Manitoba wheat which had become recognized as a high-quality product and was in good demand. A pattern of production, fitting the perceived conditions, had been developed.

The data in Table II portray the rapid development in the 20-year period between 1881 and 1901. Data for 1966 are included to give a picture of developments since that time. These show that the industry has continued to develop. Cultivated acreage tripled, but production of various crops increased to a much greater extent, reflecting improved agricultural practices and new, improved varieties of crops. Not shown in the table are new crops, such as sugar beets, corn, peas, and sunflowers, that have achieved a fairly prominent place in the production program of many Manitoba farmers in recent years.

Table II
Manitoba population and agricultural statistics

		1881	1866	1891	1901	1966
Population	No.	65,954	108,640	152,506	255,211	963,066
Cultivated land	Acres	250,416	752,571	1,232,111	3,994,560	12,446,065
Horses	No.	16,418	37,485	86,735	152,573	37,015
Cattle	"	60,281	144,685	230,696	335,778	1,151,179
Sheep	"	6,073	16,053	35,838	28,355	50,499
Swine	"	17,358	101,490	54,177	123,574	499,192
Wheat	Bu.	1,033,673	6,711,189	16,092,220	18,352,929	79,000,000
Barley	"	253,604	1,054,234	1,452,433	2,666,567	22,000,000
Oats	"	1,270,268	4,740,947	8,370,212	10,592,365	74,000,000
Rye	"	1,203	2,574	12,952	7,085	3,400,000
Flax	"	--	62,203	--	81,898	16,000,000
Potatoes	"	556,193	1,203,575	1,757,321	1,892,803	3,100,000
Apples	"	190	--	713	555	--
Hay	Tons	185,279	9,685	485,230	475,600	1,850,000
Butter	Lbs.	957,152	3,469,524	4,830,368	8,500,724	22,196,000

Census of Canada.

For 1996 the data on crop production are for 1965 and taken from the 1967 *Canada Year Book*.

Chapter VIII
Saskatchewan

In what is now the province of Saskatchewan, the earliest attempts at agriculture by settlers were in areas that today are part of the productive agricultural portion of the province. This is in direct contrast to the situation in Manitoba where, for the first 100 years, the agricultural activity was on Hudson Bay, an area that is of no agricultural consequence today. Also in Alberta, the pioneer point was outside the present agricultural area.

As in most parts of Canada, the fur traders were the agricultural pioneers. In Manitoba the English employees of the Hudson's Bay Company were the first, followed by employees of the Northwest Fur Company farther south. In Saskatchewan the Hudson's Bay Company was not the first to establish trading posts, but it was the first to record agricultural activity. Some sources give this credit to a Frenchman, La Corne, reputed to have grown wheat in 1754.[1]

The first agricultural activity of record was at the post established by the Hudson's Bay Company at Cumberland House in 1774. This is the oldest point of continuous settlement in the province. The first reference to gardening activities is found in the post journal for October 25, 1776, when it was recorded: "one man clearing some Ground for Gardening in the Spring."[2] Each year after that there were references to gardening activities without mention of the crops grown until 1784 when, on May 7, the following entry was made: "One sowing part of the Garden, but poorly of for Cabbage seed there not having above a tea Spoon full, and they very old, the rest digging a piece of Ground to sow a little barley in that was raised here last Year, by sowing a little in the Garden, which came to tolerable good Perfection and very little Inferior to that from England."[3] Then on May 11: "Sown the Barley."[3] This indicates clearly that barley was grown successfully in 1783, even though in small quantity, for it provided the seed for the 1784 sowing. At the same time, if not earlier, there also was barley production at Hudson House, further up the North Saskatchewan River. Thus in the journal of that post, on May 10, 1783, the following: "Men digging ground to sow Barley in."[4] And May 12: "Five men digging ground for sowing Barley and fencing the same."[4]

Turnips and cabbages seem to have been standard crops at all of the posts. While one might assume that potatoes would be one of the first crops grown after the establishment of a post no reference is made to this crop at Cumberland House until May 9, 1793, when: "Employed

The Trail Blazers of Canadian Agriculture

digging and sowing Cabbage seed in the garden and setting Potatoes."[5] This is not the first reference to potatoes in Saskatchewan for at the South Branch House, October 1, 1788: "'Took up my few potatoes"[6] and on May 8, 1789: "Also sowed some seed in the Garden today and some Tatoes."[6] Again on May 27: "The men at the house setting some Potatoes and one attending horse."[7] A crop was harvested for on October 12: "The men employed taking up Tatoes and minding the house."[8] Also at Manchester House there was an earlier reference for on October 23, 1792: "Sent four Men up the River to dig up Potatoes."[9] From this it seems that potatoes were a general crop and very likely were grown earlier at Cumberland House than the records indicate since Cumberland House was the staging and supply post for the posts higher up the river.

It is difficult to get an accurate estimate of the extent of potato production, but a few hints can be obtained from the various journal entries. Thus, at Cumberland House, in 1797, the following: "Oct. 9 — 3(Men) taking up Potatoes. Oct.10 — the rest employed as yesterday. Oct. 11 — 3 taking up Potatoes. Oct.12 — sent 3 men to the Old House to take up Potatoes. Oct.13 — All hands as Yesterday and finished taking up the Potatoes at the Old House — an indifferent crop. Oct.17 — 2 men taking up potatoes in the new Garden — a remarkable bad crop. Oct. 18 — Finished taking up the Potatoes."[10]

In 1798 the records show: "May 7 — 2 men finished setting potatoes in the west Garden — it has taken 4 1/2 Bushels to set it. May 12 — The rest digging and setting potatoes at the Old House there is 2 Bushels set there."[11] In 1802 the September 25 entry reported: "the other two men took up the last of the potatoes the whole of which amounts to Eighty five bushels."[12] Adding the various quantities in 1818 shows a yield of 150 kegs and the following year 390 kegs were harvested.[13]

The agricultural activities continued to increase at the posts in the north and in 1815 the first reference to the use of the plow was noted at Carlton on April 26: "Two men employed ploughing up the Gardens."[14] One might well suspect that a plow had been in use before this in view of the acreage under cultivation. The journal entry of November 1, 1814, gives an indication of the scope of operations at that time: "Produce of about 4 acres of cultivated land, that is been put by for use of the winter exclusive of what has been used since 20th August.

Potatoes	Bus. 284	Oats	Bus. 8 1/2
Turnips	" 45	Cabbage	Nr. 2000
Barley	" 66	besides some other small Vegetables."[15]	

The cultivated area was increased for on May 13, 1819: "Measured the ground in cultivation and found it to be nearly 7 acres."[16]

Cumberland House made reference to the use of the plough in 1819. Previously land had been cultivated by digging and hoeing. There is some indication that the plows in use were not particularly sturdy for

Cumberland House reported: "People employed in breaking out a piece of new ground with hoes preparatory to plowing."[17] It was bushland that was being taken under cultivation and it would be necessary to clear the bush before the plow could be used.

The use of a harrow was reported for the first time in 1815.[18]

The first reference to wheat was made at Carlton House on May 3, 1815: "this day sowed 1 Bushel of Barley, 3 1/2 pints of Wheat."[19] Cumberland House entered the wheat production field a little later when it reported on May 9, 1819: "Sowed 1/2 bushel of wheat and 2 bushels of barley today."[20] In the same year this post reported the first harvest of wheat[21] though it may be assumed that the earlier reported seedings also were productive.

That wheat continued to be grown is indicated by Franklin's report in 1820: "Carlton House... is pleasantly situated about a quarter of a mile from the river's side on the flat ground under the shelter of the high banks that bound the plains. The land is fertile, and produces, with little trouble, ample returns of wheat, barley, oats and potatoes. The ground is prepared for the reception of these vegetables, about the middle of April, and when Dr. Richardson visited this place May 10th, the blade of wheat looked strong and healthy. There were only five acres in cultivation at the period of my visit."[22] This estimate of acreage differs from that reported by the post the previous year.

The year 1819 was significant as it marked the first recorded export of agricultural products from the province. Thus, on October 12, a party was sent from Cumberland House to the Red River Settlement and: "sent with them 36 Kegs @ 10 Gall. of Potatoes, 8 Bushels of Barley, and 1 Bushel of Wheat."[23]

Even at this early date consideration was being given to quality of product, for the April 29, 1815, entry at Carlton House stated: "planting Potatoes. Planted 15 Bushels of common & 2 Bushels of a better kind called Ladys Fingers; also sowed 1 1/4 bushels of Oats."[24]

Environmental Hazards

The environmental hazards of agricultural production were encountered by the early producers from time to time. One of these hazards was summer frost and this was noted on several occasions. For example, from Alexandria, on the Assiniboine River, on July 17, 1804: "We had a frost, so hard, that it has injured many things in our garden"[25] and at Carlton House on June 22, 1815: "a sharp frost last night which Froze all the Potatoes & other tender plants."[26]

In 1802, Harmon, at Alexandria, provided a vivid description of an invasion of grasshoppers: "There are at present, in this vicinity, grasshoppers, in such prodigious numbers, as I never before saw in any place. In fair weather, between eight and ten o'clock, A.M., which is the only

The Trail Blazers of Canadian Agriculture

time of the day when many of them leave the ground, they are flying in such numbers, that they obscure the sun, like a light cloud passing over it. They also devour every thing before them, leaving scarcely a leaf on the trees, or a blade of grass on the prairies; and our potatoe tops escape not their ravages."[27]

But grasshoppers were not the only insect pests. At Carlton House, July 20, 1815, it was noted: "The insects have nearly eaten all our Cabbages & turnips owing I suppose to the dry season, its a kind of small Caterpillar & have not been troubled with any of them in former seasons."[28] In this we see also the problem of drought. This was noted on several occasions at various posts and has been a continuing hazard to agricultural production in the province. But too much rain, as well as hail, were hazards at times. For example, a letter from Cumberland House, dated August 27, 1828, reported: Our crops suffer nigh daily, from immense heavy fall of rain, with now and then, accompanied with hail, which has laid part of it flat to the ground, others became too luxuriant that it is questionable if it can ripen this season, and that, that is ripe, loses much before we can find a dry day to reap it, but our potatoes and Turnips both well."[29]

The importance of the agricultural products for sustenance of the men at the posts is noted on a number of occasions. One example is in the entry in Harmon's journal of January 21, 1805: "For nearly a month, we have subsisted on little els (else) than Potatoes! But thanks be to a kind Providence, last night two of my Men returned from the Plains with their sledges loaded with the flesh of Buffaloe:"[30] Likewise, from Carlton House, September 5 and 11, 1823: "Nothing in the House to give the people to eat but Potatoes and a little grease."[31]

It has not been possible to determine accurately when domestic animals first made their appearance in the province. The Indians had horses by the middle of the 18th century when Henday encountered them somewhere south and west of Cumberland House. The early traders soon acquired horses and used them as supplementary transportation. While there is contradiction in the various records it is certain that cattle were in the province in 1819 and may well have been there earlier. Franklin's statement regarding Cumberland House in 1820 is ambiguous with respect to animals: "The land around Cumberland House is low, but the soil, from having a considerable intermixture of limestone, is good, and capable of producing abundance of corn, and vegetables of every description. Many kinds of pot-herbs have already been brought to some perfection, and the potatoes bid fair to equal those of England. Horses feed extremely well even during the winter, and so would oxen if provided with hay, which might easily be done. Pigs also improve, but require to be kept warm in the winter."[32]

From this one might draw the conclusion that there were pigs at the post though there is no record of this in the journal. On the other hand the journal records, on August 25, 1819: "Two Horned Cattle a Bull & a Cow was brought from Swan River."[33] It is not clear where these cattle originated, but they may have been in Saskatchewan prior to that year. The cow was old enough to produce a calf for on May 16, 1820: "Skinner not being able to work is employed taking care of the Cow & Calf."[34]

There is a hint that cattle may have been at Carlton House at an earlier date for on October 29, 1819, the following entry appeared in the Cumberland House journal: "In the afternoon Wm. Ballendine, Donald Thompson, James Leith & Wenneshuse (an Indian) arrive from Carlton in a Boat with the following quantity of provisions: 120 lbs. pounded meat, 586 lbs. Fat, 320 lbs. dried meat, 1900 lbs. Buffalo meat, 2 Kegs of Salt Beef."[35] The last item could conceivably refer to salted buffalo meat, but it is rather strange that it was designated as beef if it was buffalo meat. Another item that suggests the presence of cattle at Carlton is the notation, on October 10, 1818: "Larsock attending the Stable."[36] However, for recorded fact we have to revert to the 1819 date at Cumberland House.

In any event there was considerable addition of livestock early in the 19th century. For example, Mclean reported in August 1833: "We reached Cumberland House on the 8th. Here I was cheered by the sight of extensive corn-fields, horned cattle, pigs and poultry, which gave the place more the appearance of a farm in the civilized world than a trading post in the far North-West."[37] Later that year he wrote: "We arrived at the post of Isle a la Crosse, where we were detained a day in consequence of bad weather. This post is also surrounded by cultivated fields, and I observed a few cattle; but the voice of the grunter was not heard."[38]

The various reports indicate that agriculture was being carried on successfully at a number of the fur trading posts in Saskatchewan during the last quarter of the 18th century and continued into the 19th century. The transition from agriculture as an adjunct to the fur trade to agriculture as an end in itself began late in the 19th century. Unlike the situation in Manitoba, where the arrival of the Red River settlers has been taken as the historical point of change from fur trade to agricultural settlement, no such historical landmark has been accorded the development in Saskatchewan. The agricultural development there has usually been looked upon as the normal extension from the development in Manitoba. Furthermore, most attention has been given to the significance of the railway in the development of the province. Yet, similar to the situation in Manitoba, there had been considerable agricultural development quite some time prior to the coming of the railway.

One of the earliest beginnings of this change was the move of a number of Metis from the Red River Settlement, after the first Riel disturbances, to a location on the North Saskatchewan River. While the Metis were not generally agriculturally oriented they did undertake some agricultural production. The Reverend Nisbet, who started the Prince Albert Mission among them in 1866, reported in 1871: "Farming has been carried on with great success until last year, when the crops proved a comparative failure. After the crop was reaped and stacked wet weather came and destroyed some; and thousands of blackbirds wasted a great deal of it, besides which the dry, hot weather in June checked the growth considerably. Still there were harvested 142 bushels of wheat, 73 bushels of barley, 365 of potatoes and 100 of turnips — the total value of which there is set down at £186."[39] English-speaking mixed-bloods had settled in the same general area in 1862 under the leadership of James Isbister. They also farmed to a limited extend.

The Prince Albert settlement became the nucleus for considerable development. For example, the Carlton post, where agriculture had been carried on for many years, established a farm at the Prince Albert Mission. Grant reported on this in 1872: "With regard to crops, barley and potatoes were always sure, wheat generally a success, though threatened by frosts and early droughts, and never a total failure. This year, he expects two thousand bushels of wheat from a sowing of a hundred."[40]

Another indication of the growth of the settlement is an item in the *Manitoban* of June 7, 1873: "For the Saskatchewan, a train of 18 or 20 carts and waggons passed through town on the 5th, bound for the neighborhood of Prince Albert Mission, Saskatchewan. Mr. Morrison McBeath and Mr. P. Henderson and families, Kildonan, were the owners, and were taking out their farming implements and household stuff etc. They also take out a band of cattle."[41]

The extent of development is indicated in a report by Spence, in 1877: "A few miles West of this is the new and flourishing settlement of Prince Albert, situated on the South side of the North branch of the Saskatchewan, about 45 miles below Carlton. This settlement extends for about 30 miles along the Saskatchewan, the farms fronting on the river and extending back two miles. The settlers, though principally Scotch, are composed of English, Irish, German, Norwegian, Americans and Canadians. This settlement has increased rapidly, especially within the last two years, and now numbers about 500 souls, and the people are beginning to farm extensively. Wheat sells for $2 per bushel; barley $1.50; potatoes $1.25; and butter, 37¢ per lb. Several settlers have commenced stock-raising on a large scale, and the facilities for this branch of industry are of no ordinary kind inasmuch as there is abundance of hay and pasture. As an evidence of the prosperity of the settlement, it may be mentioned that good horses, waggons, light waggons, and buggies are found everywhere. The settlers have also the most approved agricultural implements, mowers, reapers, threshing machines, etc. There are mills and stores and two schools in the settlement."[42]

"In 1874, an English gentleman, on a hunting tour, attracted by the advantages offered by this location, established, at great cost, in the settlement of Prince Albert, a steam saw and grist mill, the first in the Saskatchewan country."[43]

In the meantime, agricultural development also began at Battleford, associated with the establishment of the Territorial capital, in 1876. It is reported that in that year "Fuller also harvested the first grain grown at Battleford. He fenced and ploughed a thirty-acre plot on the tongue of land between the two rivers in the spring of 1876, and seeded it on May 11 and 12 with cereal crops and vegetables. In September he harvested about 120 bushels of wheat and 150 bushels of barley, which averaged about twenty to twenty-five bushels per acre, as well as potatoes, onions and beets."[44]

The scope of some of the operations there can be seen by that of the Finlaysons, begun in 1879: "By 1883 they had 700 acres fenced and 160 broken. They owned ten horses, thirty head of cattle, a reaper, mower, rake, five ploughs, two harrows, two seeders, three wagons, three bobsleighs and were out of debt." In 1881, the Finlayson brothers marketed and shipped, for the first time, wheat grown at Battleford.[45]

Other developments in the north were reported by the *Saskatchewan Herald* in its first issue on August 25, 1878: "A general exodus of the younger portion of the population of High Bluff and Poplar Point is predicted for this fall and next spring, the number now making preparations to leave being put at from fifty to sixty. The advance guard has already been out, and already land has been broken and other improvements made in the south branch settlement near Prince Albert."

"A large and very flourishing settlement is being made on the north bank of the South Branch of the Saskatchewan, ten miles below St. Laurent Mission. Many of the settlers are of the old families of Manitoba, while a few are from Ontario. The land is of the very finest quality, with an abundance of good wood and water. This settlement is bound to prosper, as most of those taking up land are practical men of ample means, who begin by taking in plenty of young stock, pigs, poultry, farming implements, etc.."[46]

The sequel to this story appeared in the newspaper on August 25, 1879: "A party of settlers from Manitoba, numbering nearly one hundred souls went to locate in the South Branch and Prince Albert settlements in the early part of this month."[47]

While Battleford, Prince Albert, and South Branch were the major settlements one other was started on the Carrot River, at Kinistino, in 1878. By 1884 the settlement had between 40 and 50 settlers. In 1883, some 9,000 bushels of grain were harvested from 450 acres. A total of 985 acres had been broken.[48]

It is evident that considerable development was taking place and

The Trail Blazers of Canadian Agriculture

people were moving into the area in considerable numbers. The motivation of the early settlers varied. The Metis and other mixed-bloods came to escape from the encroachment by settlers in Manitoba. What brought the next group is difficult to determine for, to modern eyes, there seemed to be little inducement. Distance to market certainly was a serious handicap. The river was the main means of transporting products out of the area though a wagon trail had developed between Manitoba and points west. Still the river was the reason for development in the north rather than in the south of the province where no river system was available. Later, the prospect of the proposed Pacific Railway to follow the Saskatchewan river west, provided an inducement. It provided stimulus to settlers and to the land speculators. However, with the change of the railway route to the south, the settlement pattern underwent a major change. The northern settlements suffered setback, with limited addition to the population, whereas the major influx of settlers followed the railroad into southern Saskatchewan.

It is difficult to get a clear picture of the population changes for the early settlement years in both Saskatchewan and Alberta as no separate statistics were kept, all data being lumped under the heading "Territories". In fact, no reliable data are available prior to 1881 when the first census was taken and, as the Territories had been subdivided into units for record purposes, it has been possible to get an approximation for each province for the human and crop statistics prior to the time when they became provinces.

An indication of what was happening is provided by records of the land office. The *Regina Leader* reported, on November 22, 1883: "During the year ended 31st October, 2,268 homesteads and 1,843 pre-emption entries covering 657,760 acres were granted. In the year 1881 the returns of seven offices show the total entries to have been 2,753 homesteads and 1,647 pre-emptions or an area of 702,354 acres. Last year when the country was swarming with settlers and almost everyone seeking a location the greatest number of entries made at any one office was 1,641 or 627 less than the Regina returns show for this year."[49]

Some indication of the immediate effect of the railroad on settlement can be gleaned from various sources. Barneby, who visited Qu'Appelle in 1885, stated: "I am informed that about one hundred settlers are already located on the lands we visited today, and it is expected that the whole district will soon be filled up."[50] The *Regina Leader* reported, on June 19, 1884, just one year after the railroad had gone through, that "between Troy and Pense exclusive of the Bell Farm and exclusive of the country around Moose Jaw there are 30,000 acres under seed."[51]

Most of the early settlers in Saskatchewan were emigrants from Manitoba and Ontario, but after the railroad came in it was not long before settlers began to come from other countries in substantial numbers. For example, in 1883, a group of Scottish crofters came and settled south of Moosomin and were followed in later years by others from Scotland. In 1885 a group of German settlers arrived and they were followed in turn by more of the same nationality. Icelanders came a year later and these were followed by Mennonites in 1891, Ukrainians in 1897, and 7,000 Doukhobors in 1889. This immigration was encouraged by the Canadian government and by the railway as both parties were interested in having the land settled.

Land Ownership a Problem

Legal ownership of land was a vexing problem for many years. The country had not been surveyed and there was no official procedure for registering ownership. All of the early settlements were located along the rivers, with all of the farms having a river frontage and the individual settlers claiming land back from this frontage. When the official survey was made, into the square section and township pattern, it conflicted with the previous system. Compromises were effected and eventually the original settlers acquired title to their holdings, but much conflict resulted from this problem.

Unfortunately, the land survey did not precede settlement in what may be called the railway era. The area of the Territories was so vast, the means of transport ahead of and away from the railroad relatively slow, so that it was physically impossible to complete the survey before settlers began to move in. The government gave preference to surveying land adjacent to the railroad which led to continued complaint from the northern settlements that they were being hampered in development.

An indication of the scope of the survey is given in an editorial report in the *Prince Albert Times* of February 1, 1884: "Nearly 120 survey parties, or 1,200 men, have been engaged in the work this summer, and they have subdivided about 1,400 townships, a greater area than was ever gone over before in one season in Canada or the United States. The section of the country which was surveyed extends from Moose Jaw and the Touchwood Hills to Calgary, and from the third base line north to the ninth correction line... The townships which have been surveyed will be ready to be placed on the market next summer...But great as the increase in the work done was, it is estimated that it will take three more years at the same rate to cover the entire northwest."[52]

The surveys did not solve all of the problems. Government land policies were a constant source of complaint and the ownership of large blocks of land by different companies also caused problems. When the Hudson's Bay Company holdings in the west were acquired by the Government of Canada, the Company was granted land surrounding its posts and also certain surveyed sections scattered throughout the Territories. The Canadian Pacific Railway Company, and later other railway companies, were granted vast tracts of land as subsidies for the construction of railways. These originally were in a belt of land, ten miles wide along the right-of-way, though later some of this was traded for land further out. In addition to the land thus granted, certain sections of land in each township were reserved for school lands, and other lands were reserved for other purposes. Settlement often preceded the surveys and sometimes settlers located on land owned by one of the companies rather than on land available for homesteading or pre-emption.

In addition to the availability of crown land, large tracts were sold to colonizing companies. A general picture of the homestead and pre-emption regulations is provided by the Dominion Lands Regulations for 1889. "Under the Dominion Lands Regulations all surveyed even-numbered sections excepting 8 and 26, in Manitoba and the North-west Territories, which have not been homesteaded, reserved to provide woodlots for settlers, or otherwise disposed of or reserved, are to be held exclusively for homesteads or pre-emptions. Homesteads may be obtained upon payment of an Office Fee of Ten Dollars, subject to the following conditions as to residence and cultivation: in the 'Mile Belt Reserve', that is the even-numbered sections lying within one mile of the Main Line or Branches of the Canadian Pacific Railway, and which are not set apart for town sites or reserves made in connection with town sites, railway stations, mounted police posts, mining and other special purposes, the homesteader shall begin actual residence upon his homestead within six months from the date of entry and shall reside upon and make the land his home for at least six months out of every twelve months for three years from the date of his homestead entry, break and prepare for crop ten acres of his homestead quarter-section; and shall within the second year crop the said ten acres, and break and prepare for crop fifteen acres additional, making twenty-five acres; and within the third year after the date of homestead entry he shall crop the said twenty-five acres and break and prepare for crop fifteen acres additional, so that within three years of his homestead entry he shall have not less than twenty-five acres cropped and fifteen acres additional broken and prepared for crop."[53]

This was the basic set of conditions, but modifications applied to land not in the Mile Belt, to town site reserves, and to coal and mineral districts. There were three options: 1) Residence within six months, unless the entry was made after the first of September in which case residence could commence the following June 1. The homesteader had to live upon and cultivate the land for at least six months for each of three years. 2) Residence within a radius of two miles of the homestead for six months of each twelve. Cultivation requirements were the same as in the Mile Belt. In addition a habitable house had to be constructed by the end of the three years. 3) Residence requirements as in option 1, cultivation of five acres, plus ten, and a habitable house had to be built before the expiration of the second year. In all cases, when the homesteader applied for his patent at the end of three years he had to provide proof that he had made permanent improvements on his land, to the aggregate value of not less than $1.50 per acre.

In addition to the homestead the homesteader could obtain an additional quarter section by pre-emption if adjoining crown land was available. The cost of pre-emption was an office fee of $10 and $2.50 per acre. North of the Canadian Pacific Railway land grant and not within 24 miles of Canadian Pacific Railway branch lines, or 12 miles of any other railway, the cost was $2.00 per acre.[54] There were modifications to these regulations from time to time but the general principle remained.

In addition to the crown land, land was available for purchase from the railway company and the Hudson's Bay Company, the former advertising 25 million acres for sale. The price varied and rebates were made if certain quantities of land were prepared for cropping.

One feature of the early railway era in Saskatchewan, more so than in Manitoba and Alberta, was the establishment of large, corporate farms and colonizing companies. Most of these came to early grief and were not a major factor in the development of the province. But they did constitute an interesting aspect of the early agricultural development. One of the first of the corporate farms, known as the Bell Farm, though officially the Qu'Appelle Valley Farming Company, was situated near Indian Head. It had a total area of some 50,000 acres. The editor of the *Regina Leader* commented optimistically on this development in 1884 : "The success of the Bell Farm disposes of all fears arising out of Summer frosts. During the last season 1,200 acres were sown with wheat, ten hundred with oats, two hundred with potatoes, roots, etc.. 23,000 bushels of wheat were grown. Of this 21,720 bushels were of the finest quality, and averaged about two pounds per bushel above the standard weight of 67 lbs., the balance, 1,300 bushels having been slightly frost bitten, though not materially injured for milling purposes. The seed from which the 1,300 bushels were raised was sown on the 1st of May too late to obtain a wet start, and to this cause is attributed the damage by frost. The cost of production is estimated at forty-two cents a bushel. They sold 10,000 bushels to Mr. Ogilvie and got 89¢ f.o.b.. They sold a good many bushels to farmers around at $1.25 for seed. The roots and potatoes were fine. We have seen the wheat, it is the very best Scotch fyfe No. 1."[55]

A description of the operations is provided by Black: "From his actual report presented in January, 1884, we learn that during the summer and fall of 1882, when active operations commenced, 2,700 acres of land were broken. The land seeded this spring yielded an average of twenty bushels to the acre. In 1884 about six thousand acres were under crop. In 1883 a thirty thousand bushel granary was built, together with two large barrack cottages for the accommodation of the men at the main station, buildings for the storage of implements, a blacksmith shop, a horse infirmary and twenty-two cottages with their outbuildings costing eight hundred dollars each. Fencing, tree-planting and other improvements on an ambitious scale also received attention. During the first two years of its history, the company spent approximately $250,000. Various means were taken to reduce by cooperative methods the expense of the enterprise. The whole tract was divided into smaller farms. Two-thirds of each of these, as they were broken, were cropped each year, and one third summer-fallowed."[56]

One cannot but be impressed with the scope of this undertaking and the high degree of management that made the operation possible. Assembling the necessary manpower, horses, and equipment to break 2,700 acres the first year, would be a considerable undertaking in itself to say nothing of providing housing, food, and feed. An equal amount of land was broken the next year as well as seeding and harvesting the crop on the original 2,700 acres, in addition to carrying on a major building program. But despite what appeared to be a favorable beginning the operation was not a financial success.

Another corporate farm was the Canada Agricultural, Coal and Colonization Company with holdings of 110,000 acres in 11 different farms strung along a frontage of 50 miles on the Canadian Pacific Railway. The original farm was that of Sir John Lister Hay, at Balgonie, east of Regina. An idea of the scope of operation is given in a report in 1889 when 5,000 acres were under crop at all farms, some of it being in wheat, but most of it in oats and barley as feed for the large number of animals on the farms.

"Last year nearly 8,000 cattle were bought from the now defunct Powder River Cattle Co. — 100 Polled Angus and Galloway bulls, 12 stallions and 600 brood mares. There is now accommodation provided for 500 horses and 6,000 cattle, shedding for over 30,000 sheep and 3,000 hogs."[57]

None of the corporate farms was financially successful and all disappeared without having had much effect in bringing settlers into the province.

The colonizing companies were more directly concerned with bringing settlers to the land. The basis on which they operated is described by Black. "Any person or company satisfying the government of good faith and financial stability might obtain, for colonization purposes, an unsettled tract of land anywhere north of the main line of the Canadian Pacific Railroad, not being within twenty-four miles of that road or any of its branches, nor within twelve miles of any yet projected line of railway. The even numbered sections were held for homestead and preemption purposes, but the odd numbered sections would become the property of the colonization company, on payment of two dollars per acre in five equal instalments. The company would also pay five cents an acre for the survey of the land purchased, and interest of six per centum would be charged on all overdue payments."

"The contract into which the colonization company entered with the Government required that within five years the company's reserve should be colonized by placing two settlers on each odd numbered section, and also two settlers on each of the free homestead sections. When such colonization was completed the company was to be allowed a rebate of $120 for each bona fide settler. On the expiration of the five years, if all conditions had been fulfilled, such further rebate would be granted as would reduce the purchase price to one dollar per acre. If, however, the full number of settlers required by regulations had not been placed upon the land in conformity with the official regulations, the company was to forfeit $160 for each settler fewer than the required number."[58]

Quite a number of companies were formed. An example of one was the Land and Colonization Company of Canada, organized by Rev. A.J. Bray, of Montreal, with a capital of $5 million. This company was granted one million acres of land alongside the Prince Albert settlement.[59]

The editor of the *Prince Albert Times* gave his evaluation of these companies in 1886: "We have never had any faith in colonization societies. They have, with few exceptions, been a blight on every district where they have taken root. Their promoters claim to have the best interest of the country at heart, but it is impossible not to see that their principal object is to ultimately, if not immediately, make money out of the settlers."[60] On the whole the companies failed to meet their obligations though they did manage to bring a number of settlers into the province.

Transportation Problems

While the coming of the railroad made transportation of farm products to market easier it also generated a number of problems in addition to the land problems already mentioned. From the standpoint of transportation two factors were important. One of these was that the elevator companies moved in and established warehouses and elevators for handling the grain, and the railroad collaborated with them to the extent that it would not supply grain cars to the individual farmers. The argument used was that, with storage and handling facilities, the elevator companies could fill cars more quickly than the individual farmers and thus, could maximize use of the cars. The individual farmer might well require several days to haul his grain to fill a car. There was also the tendency for the railway company to give preference to the farmers and elevator companies in Manitoba where the shipping distance to market was shorter and the turnaround time for the cars somewhat less. The elevator companies came in for their share of criticism from the farmers, being accused of all sorts of shady practices,

The Trail Blazers of Canadian Agriculture

including short weights and excessive dockages.

These various grievances led eventually to the creation of farmer organizations, formed for the purpose of approaching the government with their problems. Eventually the government intervened and regulated both the railway and the elevator companies. One immediate result was that the railway had to provide loading platforms so that the farmers could load directly into the cars, and also gave the individual farmers equal right to cars.

The 1884 freight rate structure shows the shipping costs for farmers at various locations and indicates clearly the cost advantage of those in the Red River valley area. The *Regina Leader*, apparently a supporter of the railroad company, reported, in 1884: "C.P.R. wheat rates to Port Arthur have been reduced and are now as follows: from Regina 40¢ per 100 lbs.; from Virden 36¢; from Winnipeg 28¢; from Manitou 30¢. There is also a reduction of the rates to the boundary line at Gretna and Emerson by the rail route. The C.P.R. is wide awake."[61]

The elevator network that developed gradually was made up of single elevators owned by individuals or by small companies, as well as groups of elevators owned by larger companies. The number of elevators and warehouses grew as the railroad was extended into new areas and by 1900 there were 69 elevators and 24 grain warehouses in the province.[62]

Because of the time at which agricultural settlement took place in Saskatchewan the settlers benefitted from well developed farming equipment. Just when the first threshing machine was introduced into the province is not clear, but by 1877 Prince Albert had threshing machines and grist mills.[63] The Lieutenant Governor imported a thresher to Battleford in 1878,[64] very likely the first one in that settlement. In 1880. Mr. F. McKay of Prince Albert, imported a Waterous portable engine and a thresher and separator.[65] This may well have been the first steam-powered threshing outfit in the province, the earlier ones had all been horse driven. It was not the first steam engine as there was a steam grist mill at Prince Albert in 1874.[66]

While the early threshers were a great improvement over the flail they had some limitations. In one case it was reported that the horses were too small and could not generate enough power for the machine to operate efficiently. At Prince Albert it was reported, one fall, that the threshers were hard at work but, if there was any fault, it was that the crop was too heavy.

Not all of the settlers had a full line of modern equipment. Elkington, who farmed for a short time at Qu'Appelle, described his seeding procedure in 1850 as follows: "As soon as it was fit I began to sow wheat on my land, doing a small piece at a time and then plowing it under; in this way, one of the best methods of sowing, but apt to make the crop rather late, I put in twelve acres of wheat and twelve acres of oats."[67]

That fall he and a neighbor bought a binder in partnership. "We worked three horses on the binder, one of us driving, and the other following and setting up the sheaves."[68]

Amongst the novel equipment brought in was a steam plow tested at Indian Head in 1883.[69] There is no evidence that this was a successful venture at that time.

According to Hawkes,[70] the press drill was unknown in 1885 as it was put on the market in Winnipeg in 1888. He suggested that it arrived just in time as there was need for a machine that could put the seed a little deeper into the ground.

Fencing was another function that benefitted from new developments. The early settlers had to depend on wooden rails for their fences and this created problems once the settlements moved away from the wooded river valleys. Fortunately, a new fencing material became available, namely, barbed wire. It was advertised in the *Saskatchewan Herald* in 1830 and may have been available prior to that time. It became the almost universal fencing material within a few years. Elkington, in 1891, noted that "barbed wire was about the cheapest in the end and the most serviceable that could be got."[71] He bought a mile of it and put up his fence with posts twelve feet apart, two strands of barbed wire, and a rail on top to show up in the dark. This latter practice may have been desirable but was not generally used.

The early settlers in Manitoba had been beset by numerous environmental hazards to their crops and the settlers in Saskatchewan did not escape the same. Frosts, both late spring and early fall, caused crop damage of varying degrees from time to time. Grant, in reporting on his visit to Carlton in 1872, mentioned that at Prince Albert "barley and potatoes were always sure, wheat generally a success, though threatened by frosts and early droughts, and never a total failure."[72] Wright noted that "in 1883 the prairie credo of progress was muted when, on September 7, Arctic air, flowing down from the tundra and Hudson Bay blighted with frost the ripening crops."[73]

A realistic picture of the vagaries of the weather and its effect is given by Alexander Kindred, located south of Wolesly: "(in 1885) we had only ten bushels (per acre) of very badly frosted wheat. I took some to Indian Head and traded it for flour, shorts, and bran. I had no money to pay expenses... In 1886 we had 80 acres under crop. Not a drop of rain fell from the time it went in until it was harvested. I sowed 124 bushels and threshed 54. In 1888, we began to think we could not grow wheat in this country. I had now 120 to 125 acres under cultivation. We put in 25 acres of wheat, 10 to 15 of oats, and let the rest go back into prairie. That year we got 35 bushels (of wheat) to the acre! So we went to work and ploughed up again. The next year the wheat headed out two inches high. Not a drop of rain fell that whole season until fall. We summerfallowed that year (1889) for the first time, and to show optimism, we put in, in

The Trail Blazers of Canadian Agriculture

1890, every acre we could. We had wheat standing to the chin, but on the 8th of July a hailstorm destroyed absolutely everything. My hair turned grey that night."[74] Kindred was not the only one to suffer as drought was widespread in several years of the decade and was referred to on numerous occasions in the newspapers. Hailstorms, too, were of frequent occurrence.

There wasn't much that the farmer could do to protect himself from these hazards though Hawkes gives a description of the use of smoke smudges as a protection against fall frosts. "When the farmer was in fear of frost he would place stable manure alongside the edge of the crop, and then watch the thermometer closely. When the temperature fell to a certain point he would fire his smudges, or some of them. The efficiency or otherwise of the smudge, was often in those days a matter for serious, and sometimes heated discussion. There is no doubt that in some cases this smudge screen was effective, because wheat so protected was sometimes free from frost while that on an adjacent farm would be affected."[75] Regardless of whether it was effective it could be used only on relatively small acreages and could not be a major factor in protection. The development of wheat that would mature in the short growing season was a partial solution, but did not materialize until a later date.

Spring frosts also caused trouble. In 1891 the grain was just coming up nicely at Qu'Appelle when a severe frost struck, destroying some of the wheat and forcing many to reseed their oats.[76]

Weeds were a problem at an early date and the Territorial Council gave them consideration as early as 1883 when it was reported, regarding the Legislative Session, that: "Mr. Wm. White, of Regina, has been attentive. He has wrestled the noxious weed, and the Canada thistle has felt him at its throat."[77] The noxious weeds listed at that time, in addition to the thistle, were wild mustard, cockle, and wild oats.[78] In 1892, French weed, pigweed and lamb's quarter, and tumble weed were added to the list.[79]

The grasshopper scourge was not unknown, but it apparently was not of major consequence during the first years of settlement. Other insects caused minor damage. Caterpillars denuding cabbages and turnips were reported as early as 1815[80] and the cabbage worm in 1880.[81]

Two other enemies of crop production, the gopher and the blackbird, were reported from time to time. The former was ever present causing damage throughout the growing season whereas the latter was most destructive when the crop was ripe. There was no fully effective control for either of these pests, but the damage inflicted by blackbirds decreased as the acreage in crop increased.

No special mention was made of any crop diseases in the newspapers of the early settlement years though smut of wheat was referred to somewhat incidentally in the early development at Prince Albert. That this disease was prevalent may be inferred from a statement in the *Prince Albert Times*, in 1886, following the introduction of new wheat as a relief measure after a crop failure. "A great number of farmers are under the impression now that they have new seed wheat, they will not require to use any preventative means against smut, but from conversations we have had with leading farmers we would advise them to use blue stone, lime or other dips as usual."[82] From this it is clear that treatment of seed for control of smut had been a common practice.

The economics of crop production was a matter for early concern and there were differences of opinion regarding the profitability of grain growing. In the item about the Bell Farm it would appear that money could be made, with the cost of production at 42¢ per bushel and returns ranging from 89¢ to $1.25 per bushel.[83] An 1884 estimate by a farmer stated: "Eight dollars an acre will plow and seed and harvest and thresh an acre of wheat. This acre at the lowest will produce 20 bushels and 60¢ a bushel can be had at the store. This is a low estimate. This leaves $4.00 for the farmer per acre and no such margin can be had in Ontario."[84] A less optimistic analysis was made by another farmer in a letter to the editor; in 1885, under the heading "Cost of raising wheat — 25 bushels per acre."

One fourth the breaking	$1.00
Backsetting or second ploughing	2.75
Harrowing twice	.75
Seeding with drill seeder	1.00
Seed 1.5 bushels at $1.25	1.88
Harrowing once	37
Rolling once	.25
	$8.00
Cutting, binding and stooking	$1.55
Hauling and stacking	2.00
	$3.55
Threshing 25 bus. @8¢	$2.00
Cleaning and hauling four miles to market	1.80
	$3.80
Grand total	$15.35
25 bushels at 60¢	15.00

The farmer who sells his wheat at 60¢ sells below cost."[85]

It may be noted that no costs were shown for taxes, cost of land, and other items that are included in modern cost accounting.

Grain production took precedence among the Saskatchewan settlers though some livestock production did occur. In contrast to the situation that developed in Alberta, ranching did not play a major role in the early settlement of Saskatchewan. For example, a correspondent to the *Nor West Farmer*, in 1889, commented: "It is a matter of regret that the rich grazing grounds of Southern Assiniboia are not utilized by ranchmen...and yet with the exception of small heards (sic) at great intervals, there is no stock east of the Cypress Hills."[86] There was some ranching farther west for in the same paper another correspondent reported that the spring round-up had been completed with satisfactory results. About 9,000 cattle had been seen.[87]

In addition to beef cattle some dairy production developed and, in 1889, it was reported that the Grenfell cheese factory had produced 60,000 pounds of cheese the previous season. At that time an effort was being made to establish a cooperative cheese factory at Langenburg.[88] Encouragement for mixed farming took several forms. For example, at the Territorial Exhibition at Regina, in 1895, there was a special class for a composite exhibit of two bushels each of wheat, barley, oats, and peas; one half bushel of flax; and one male and two females of cattle, pigs, and sheep.[89]

The early settlers were venturesome. As early as June 1879 the *Herald* reported that "A number of experiments are being tried with cultivated grasses. The Governor has sown some Hungarian grass, and Inspector Walker and James Pruce have put down enough timothy and clover seed to fairly test their adaptability to this soil and climate."[90] A year later it reported that several varieties of new wheat had been introduced, including Russian and red fern, reportedly popular in the east. It was stated further that improved varieties of fall wheat also would be sown that year to test their adaptability.[91] It was hoped that fall wheat could be used to overcome the frost hazard to spring sown wheat. This did not prove to be successful. One rather interesting experiment was the testing of three varieties of sugar beets at Whitewood, in 1890, with the thought that a sugar beet industry might be developed.[92] Nothing came of this.

Individual farmers were not alone in such activities. They organized agricultural societies as had been done earlier in the other provinces. In Prince Albert, as early as 1879, it was proposed to organize both an Agricultural Society and a Rifle Association.[93] The Lorne Agricultural Society, was formed at Regina on March 30, 1884,[94] and it held its first fair in October of that year with 986 entries and $1500 in prizes.[95] Fairs also were held at Moose Jaw, Indian Head and Prince Albert that year. These were rather remarkable achievements when one considers that at all of these points, except Prince Albert, settlement had commenced only a year or two before.

Grants to Societies

In 1886 the Territorial council passed an ordinance to incorporate agricultural societies in the Northwest Territories.[96] To encourage formation of such societies provision was made for supporting grants, not to exceed the amount subscribed by the members of any society. This ordinance continued in force, with amendments, for many years. Partly because of this support there was a fairly steady increase in the number of societies and a consequent increase in the number of fairs held.

The fairs and exhibitions in Saskatchewan followed essentially the same pattern as those in Manitoba and so will not be discussed in detail. In addition to the local fairs, a new development was a major territorial exhibition held in Regina in 1895. A significant purpose of this exhibition was to promote immigration and thus the federal government gave it strong support with a grant of $25,000. The railway also was directly interested in increasing settlement and it offered free transportation for exhibits to and from the exhibition. The Territorial Council promised $10,000 and the Regina Town Council a like amount.[97]

Agricultural societies were not the only farmer organizations formed. At what might be considered to be a more political level, a meeting was arranged for February 1884 for the formation of a Farmer's Union. This led to the formation of the Assiniboia Farmers' Association "to promote the real interest of the farmers of Assiniboia."[98]

In 1890 legal provision was made for the formation of Farmers' Institutes, with somewhat the same objectives as the agricultural societies, and in 1891 the Dairymen's Association of the North West Territories was incorporated. This preceded the Western Stock Growers' Association by five years.[99] All of these organizations were concerned with developing the agricultural potential of the province.

The newspapers tried to do their bit in the interest of better farming by publishing articles and general information on agricultural topics. Unfortunately, as in Manitoba, the material for many of the articles was drawn from sources where conditions were quite different and, consequently, was not always fully applicable to the new territory. A sampling of titles will provide a picture of subject matter dealt with: deep ploughing, nitrate of soda for wheat, fowl hints, poultry on the farm, horse breeding, pulling at the halter, care for self sucking cows, kicking cows cure them, distinguishing characters of different breeds of cattle.

A new element in the development of improved agricultural practices was the Experimental Farms system established by the federal government in 1885. The first director, William Saunders, travelled through the west to familiarize himself with the conditions and to select sites for the first experimental farms. He also conducted meetings, instituted tests of varieties of grain by providing samples of seed to individual farmers, and made arrangements for the distribution of publications. Of special importance to Saskatchewan was the establishment of the Experimental Farm at Indian Head, in 1887, one of five original farms across Canada.

Judging by letters to the editor and by editorials it is evident that there was not whole-hearted support from farmers and newspaper editors for such a farm, though the need for new information was evident and individual farmers and the agricultural societies had conducted some tests. At the same time there was competition among communities for the farm, with Regina making a strong play for it.

The first newspaper reports of the activities at the farm appeared on April 3, 1888. "Mr. Angus McKay, the manager of the experimental farm at Indian Head has been east to purchase stock and implements. The latter he bought in Winnipeg. He gave the following particulars to a Winnipeg man. The farm consists of 680 acres, partly light and partly heavy soil. Though it has been previously cropped it is in bad shape, and he proposes to summerfallow most of it so as to start fair next year."[100] Then on May 8: "Work on the experimental farm is going ahead. Many kinds of grain have been sown. The planting of fruit trees is also being carried on vigorously."[101] Despite the early apparent lack of enthusiasm for the farm the farmers came to realize the contribution that it made to the development of agricultural practices in the province.

Legislation Enacted

Direct government involvement in matters affecting agriculture became evident in the early years of the Territorial government. The early legislation affecting agriculture in the province was similar to that enacted in Manitoba and reflects the problems of concern. The most important acts are listed but without indicating reenactments or amendments that were made from time to time.[102]

1877
- An ordinance for the prevention of prairie and forest fires.

1878
- An ordinance respecting fences. This defined a legal fence.
- An ordinance respecting the marking of stock. This was the first brand act.
- An ordinance respecting stallions. This prohibited the running at large of stallions within ten miles of settlement.
- An ordinance respecting poisons. This prohibited the use of strychnine for destroying wolves except under certain conditions.

1879
- An ordinance respecting the registration of deeds and other instruments relating to lands in the Territories.

1881
- An ordinance respecting trespassing and stray animals.
- An ordinance respecting the protection of sheep. This made it lawful to destroy dogs that were worrying sheep.
- An ordinance respecting theft of animals.
- An ordinance respecting bulls. No bull one year or older to be at large between February 1 and June 1.

1883
- An ordinance respecting infectious diseases of domestic animals (specifically glanders in horses) and the destruction of such animals.

- An ordinance to enforce the destruction of Canada thistle and other noxious weeds, wild mustard, cockle, and wild oats.

1884
- An ordinance respecting the herding of animals.
- An ordinance to encourage the planting of forest trees.

1886
- An ordinance to incorporate Agricultural Societies. This included provision for grants to not exceed the amounts subscribed by members of the societies.

1887
- An ordinance to prevent the pollution of running streams. This prohibited the disposal of stable manure, night soil, carcasses or any filthy matter along the bank or in the water. The bank included the land within 50 feet of the high water mark.

1889
- An ordinance to provide for the incorporation of butter and cheese manufacturing associations.

1890
- An ordinance respecting the establishment of Farmers' Institutes. This provided for a maximum grant of $50 to each Institute.

1891
- An ordinance to authorize the formation of an association under the name of the Dairymen's Association of the North West Territories.

1892
- An ordinance respecting the veterinary profession. This provided for the establishment of the Veterinary Professional Association of the North West Territories.
- A new ordinance adding French weed, tumble weed, and pig weed or lamb's quarter to the list of noxious weeds.

1894
- An ordinance respecting the formation of irrigation districts.

1895
- An ordinance respecting stock injured by railway trains.
- An ordinance respecting threshers' liens.

1896
- An ordinance to incorporate the Western Stock Growers' Association.

1897
- An ordinance respecting the Department of Agriculture. This established a Department of Agriculture.

The Trail Blazers of Canadian Agriculture

The financial support for agriculture is indicated by the budget provisions for 1897:

Agricultural Societies	$ 3,500
Destruction of wolves	3,000
Destruction of noxious weeds	2,000
Special assistance to the dairy industry	1,000
Grant to the North West Dairy Association	750
Collection of statistics	1,800

Official agricultural statistics were first collected in 1881, but did not include a record of the number of livestock. The data for 1881 and some subsequent years are shown in Table I. These provide an indication of the rate of agricultural development. If we take 1866, the year the Prince Albert settlement was begun, as the beginning of agricultural settlement, as distinct from the fur trading posts, it is evident that the 15-year period to 1881 had brought about considerable development, with 14,000 acres of cultivated land. Then, with the coming of the railway, real expansion took place.

By 1901 it was clear that agriculture was developing into an important industry and the pioneering phase was ending though the major thrust of settlement was still to come in the early years of the 20th century. The province, still a part of the Northwest Territories and not yet a separate political entity, was well on its way to becoming the major wheat producer that it now is. Oat production was relatively great, reflecting the fact that this grain was needed as feed for the horses used for farm work, as well as for other classes of stock. Some crops, not shown in the table, were of early interest but never became significant. These included hemp, flax for fibre, corn, buckwheat, peas, and root crops.

The data for 1966, included in Table I, show that the early promise of agricultural development has been fulfilled despite the setbacks that have occurred from time to time by drought, frost, and insect pests.

Table I
Population and agricultural statistics for Saskatchewan

		1881	1891	1901	1966
Population	No.	20,273	40,206	91,279	955,344
Cultivated land	Acres	13,901	--	25,188	45,468,776
Horses	No.	--	27,918	83,461	74,674
Cattle	"	--	81,933	217,053	2,397,979
Sheep	"	--	39,151	73,097	127,799
Swine	"	--	11,045	27,743	488,173
Wheat	bu.	69,007	1,697,071	4,306,091	400,000,000
Barley	"	23,741	125,161	187,211	65,000,000
Oats	"	26,240	1,050,530	2,270,057	94,000,000
Rye	"	240	1,299	12,883	7,300,000
Flax	"	--	709	2,430	7,300,000
Potatoes	"	50,357	342,715	692,254	970,000
Hay	Tons	12,736	108,948	251,992	2,075,000
Butter	Lbs.	--	1,469,451	2,236,535	22,755,000

Data for the years 1881, 1891 and 1901 are from the census reports for these years.

The first six items for 1966 are from the census report for that year. The remaining items are for 1965 from the 1967 *Canada Year Book*.

The Trail Blazers of Canadian Agriculture

Chapter IX
Alberta

The area that now is the Province of Alberta was the last prairie area to be explored and exploited. There still seems to be some dispute as to who the first non-native was to set foot in the area. One claim is that de Nireville established a fort (La Jonquirere) at a site close to the present city of Calgary in 1751, but there is no sound evidence to support this claim. Henday ventured into the area between the North Saskatchewan and the Red Deer rivers in 1754-55. However, there is nothing to indicate that there was any agricultural activity in either case.

The movement west followed the northern rivers as these led to the rich fur country whereas the areas to the south, while useful as a source of meat supply from the buffalo, were less valuable as a source of high-quality furs. As well, the Indians of the southern prairies were relatively unfriendly and early attempts at establishing posts in the south had not been successful. Early agricultural activity occurred at the trading posts; therefore, the earliest such activity in the province was in the north. The first agricultural activity in the southern part of the province came almost a hundred years later.

As far as can be determined with accuracy, the first agriculture in the province occurred on the Athabaska River about 40 miles south of Lake Athabaska where Peter Pond established a post in 1778. Swindlehurst refers to this as follows: "Long before the buffalo had ceased to roam the plains, Fur Trader Peter Pond at his isolated post near Lake Athabaska, was raising vegetables in his little garden. That was in 1779. He was the first to establish a post on the Athabaska River and the first white man, in the land that is now Alberta, to cultivate the soil."[1]

That Pond was a good gardener is evident from the report of Alexander Mackenzie who wrote: "In the fall of the year 1787, when first arrived at Athabaska, Mr. Pond was settled on the banks of Elk River, where he remained for three years, and had formed as fine a kitchen garden as I ever saw in Canada."[2]

Though this was the beginning of agriculture in Alberta, it was in an area that is not agriculturally important today. But there was early agriculture at another site on the edge of what today is an important agricultural area. This was at a post on the Peace River a few miles above Boyer River, near present-day Fort Vermilion. As Mackenzie reported: "In the summer of 1786, a small spot was cleared at the Old Establishment, which is situated on a bank thirty feet above the level

The Trail Blazers of Canadian Agriculture

of the river, and was sown with turnips, carrots, and parsnips. The first grew to a large size, and the others thrived well. An experiment was also made with potatoes and cabbages, the former were successful, but for want of care the latter failed. The next winter the person who had undertaken this cultivation, suffered the potatoes, which he had collected for seed, to catch the frost, and none had been since brought to this place. There is not the least doubt but the soil would be very productive, if a proper attention was given to its preparation."[3] The above is the first reference to the production of potatoes in Alberta though one may surmise that Pond had had some in his garden.

Moving further south, to what today is an area of major agricultural significance, we note the establishment of Fort Edmonton in 1795. There is evidence that a garden was established very early in the history of this post for Sutherland's journal entry of October 12, 1796 states: "The men employed getting hay home. Two men at the pit-saw and the rest cutting down stockades to enlarge the yard and gardens etc.."[4] The garden was not mentioned in 1797 but on May 4, 1798: "the rest planted a few potatoes"[5] and on May 7: "All the rest digging the ground and planting potatoes."[6]

The importance of the garden is indicated by further entries in the journal. Thus, on September 21, 1798:, "and the rest finished a cellar for preserving the garden stuff in."[7] This was followed by the October 3 notation: "The rest employed bringing home hay and took up the garden stuff."[8] The full value of the garden produce is noted in a later entry: "but as to provisions we have been in a starving condition ever since our arrival at this place, not being able to serve half allowances and there is not likelihood, of its mending and had it not been for the garden stuff it would have been worse."[9]

The post called Edmonton was moved several times before becoming established permanently at the present site of the city of Edmonton in 1812. Agriculture was carried on at all of the sites.

Meanwhile there were new developments in the Peace River country where a substantial post had been established by McLeod at the forks of the Smoky River. This is described in James Mackenzie's journal as follows: "At Mr. McLeod's fort, the men's houses are better arranged than the Bourgeois' houses here. The fort is built with five bastions, courtyards are made everywhere, a spacious garden is made around the fort, a well, a powder house and even a — house are made in this garden."[10] That this garden continued to be significant is indicated at a later date when Wallace, from Dunvegan, wrote: "and McLeod's great establishment at Smoky Forks was vacant at least during the summer. The 'Spacious garden', however, which in 1800 had excited the distant envy of James Mackenzie at Fort Chipewyan, was of sufficient importance to have men sent from Dunvegan to attend to it during the season."[11]

The Dunvegan post was established in 1806 and immediately started gardening activities as noted in journal entries the same year. "May 14 — Cadotte, Baptiste and ourselves hoed part of the ground in the garden. May 15 — F. Goedike made a hot-bed in which cabbage seeds are to be sowed. May 16 — J. Hoole arrived from the Forks where he had left old Paquette making a garden. May 17 — We cut up two kegs of potatoes to plant them. F. Goedike sowed some cabbage seeds in the hot-bed. May 19 — Old Paquette arrived from the Forks where he sowed several kinds of seeds, and planted three kegs of potatoes. We planted one keg of potatoes. May 21 — F. Goedike and Landrie sowed barley."[12] This is the first reference to barley in Alberta.

Harmon spent several years at Dunvegan and several items from his journal provide a good picture of the conditions there in the years 1808-1810. October 10, 1808: "Here our principal food will be the Flesh of Buffaloes, Moose, Red Deer, & Bears. We also have a tolerable Kitchen Garden, therefore we have what would make the most of People contented."[13] May 6, 1809: "The Plains around us are on fire. Planted Nine Bushels of Potatoes & sewed the most of our Garden Seeds."[14] July 21, 1809: "We have cut down our barley, which is I think the finest I ever saw in any Country. In short the soil of the Points of land along this River is excellent."[15] October 6: "As the weather begins to be cold, we have taken our vegetables out of the Ground, which we have found to have been very productive."[16]

The following year was even better as shown by his journal of October 3, 1810: "We have taken our Potatoes out of the ground and find that nine Bushels planted on the 10th of May have produced upwards of two hundred and fifty."[17] These results are somewhat surprising in view of the fact that on June 23 he had noted: "The last night was so cold that the tops of our potatoes are frozen."[18]

In the meantime activity was going on in the area east of Edmonton where Alexander Henry was established at Fort Vermilion at the mouth of the Vermilion River. In 1809 he reported on some of his activities in his journal: "Oct.5. I set a party gathering turnips and potatoes, as we had a frost last night, and ice in the small ponds. Oct.6. Gathered all my turnips — about 50 bushels, very large and of an excellent quality. Oct.9. Finished gathering potatoes — 80 bushels, but small and watery. The hard dry soil is unfavorable for them, being in the plains, where no wood has grown."[19] This last remark is interesting as it indicates that this site was in prairie country whereas most of the posts were located in the river bottoms where there was some tree and shrub growth.

The following year Henry had moved to White Earth House and here we learn for the first time that poultry were being maintained, though there is no record of when they first were brought into the area. Some journal entries tell of his activities. "June 13. One of my hens laid an egg; she was molested on leaving Fort Vermilion, when she had been

The Trail Blazers of Canadian Agriculture

sitting two days on 13 eggs. Some of our potatoes began to appear in the open field. June 16. I sent Hamel to sow turnips and radishes in the woods. I shot a few pigeons in our barley field, as they pluck it up by the roots and devour it.[20] June 30. My yellow hen, that has not laid since last winter, laid an egg this morning and looks as if she would continue; if so I shall have two eggs a day.[21] Aug.7. Went down to see our hay makers; they have one stack of about 400 bundles made, and 43 cocks ready to stack, but it is three miles from the house, and it will be tedious business to haul it home.[22] Aug.18. I put twelve eggs under my yellow hen."[23]

His success with increase in his poultry was not great for on August 26: "This morning I broke the 5 remaining eggs under a hen that had been sitting since the 1st inst., finding 4 dead chickens and 1 egg bad; thus of 11 eggs placed under her the 2nd, I have but 2 chickens."[24] Possibly better success with his hatchery operations was suggested in a later entry on September 23: "My black hen began to hatch her brood."[25]

Back at Edmonton developments were under way as we find the first record of wheat growing in the province in the journal of October 18, 1811: "Thrashed a little Wheat which I grew in our Garden, from three Quarts of Seed given me by Mr. Hughes we have only obtained four Gallons, and the Grain of this is so small and unripe as to be thought unfit for seed; but the past season has been particularly unfavorable for Gardens, the spring was uncommonly late, the first part of the summer remarkably dry, the latter as much too Rainy, which, with the earliness of the Fall has prevented the Barley from ripening; so that notwithstanding the Failure of this first Attempt I think there is little room for doubt that, in a favorable Season, Wheat will grow here in perfection. Of Potatoes we have about 100 bushels of an inferior Quality."[26]

Though this is the first actual reference to wheat growing in Alberta the fact that the seed was obtained from Hughes of the adjacent Northwest Company post, suggests that it might have been grown there the previous year and that wheat may have been grown there even earlier. This record of wheat production is earlier than anything noted for Saskatchewan. Despite the fact that the only varieties available had a long growing season, and were subject to frost damage, the growing of wheat continued, though not always with marked success.

The purpose of growing wheat was to make a supply of local flour available to overcome the need for the long and expensive transportation from the east. The first reference to flour being made from wheat grown in the province is from Fort Vermilion, on the Peace River, in 1826. A journal entry of October 4 includes a letter from Edmonton, dated September 23: "I have brought in a Patent Mill which I have reshipped for your place, I had it examined here & find it perfect & one of your men Catafore is instructed in the management of it. I hope it will save the freight of 7 or 8 — annually for this Dist. If convenient I should like you to send down a small bag of Peace River Flour as a specimen."[27]

The journal entry of October 5 gives the results of this: "Sent a sample of our Flour to Mr. Stewart according to desire But finding the Mill very hard to turn I am afraid it will not answer the purpose it is intended, of supplying the District with Flour, even if we have the grain to grind, unless we have a person who could make machinery for it to be turned by a Horse."[28]

Certainly there was a real question of whether they would have grain to grind. That year the yield of wheat was 7 1/2 kegs[29] but the following year: "our Wheat has failed entirely & the little Barley collected will scarcely be good for seed."[30]

With the move of the Edmonton post to its final site in 1812, agricultural production increased gradually. Hand cultivation, i.e. digging and/or hoeing the ground in preparation for seeding, was superseded by plowing and harrowing. The use of the plow was recorded for the first time on April 24, 1815:[31] "Two men ploughing, one man cutting Potatoes for seed" and on the 27th: "five men ploughing, harrowing, and setting Potatoes." The next day: "Finished setting eighty Gall of Potatoes and sowing twenty Gallons of Barley." But they were not yet through as shown by additional entries: "April 29. Sent three Men to plough and sow Barley at Old Edmonton House... James Folster cutting Potatoes for setting and the rest of the Men breaking out ground in the new tract of the Garden. Sowed twenty-five Gallons of Barley. May 1. Sent three Men to plough ground and sow Barley at the Old House the rest of the Men digging new ground in the part of the Garden lately taken in — Sowed fifteen Gallons of Barley. May 3. Five Men employed ploughing and setting Potatoes in the Garden."

Potato production continued to be an important activity though yield varied from year to year: 1821 — 400 kegs; 1822 — 350 kegs; 1823 — 700 kegs, 1826 — 1400 kegs; 1832 — 2827 kegs.[32]

The successful development of agriculture is indicated by Stuart in a visit to Edmonton in 1827: "Both at that place (Carlton) and Edmonton the crop of Potatoes, Barley and Wheat was very abundant and if a scarcity of Provisions again prevail it will not be in consequence of Buffaloe not being abundant."[33]

Crop production in the early years was subjected to most of the hazards that still exist, but there were others that no longer are serious. Among the latter, from Dunvegan, June 1, 1835: "The weather very dry so far which is not favorable to our fields. The Pigeons are numerous also and do no good to our grain. Dry weather continues to be a hazard but wild pigeons are no more."[34] Pigeons also caused trouble at other posts for at White Earth House, in 1810 Henry reported: "I shot a few pigeons in our barley field, as they pluck it up by the roots and devour

The Trail Blazers of Canadian Agriculture

it."[35]

At Fort Vermilion a different pest was noted on September 6, 1825: "Got part of our Wheat cut — the grain is very indifferent & our field will not yield much besides the small striped Squirrels destroy a great deal."[36] At Edmonton these animals also were a pest for we see on November 22, 1815: "Robert Kirkness winnowed all our barley which amounts to fifty five bushels the produce of three bushels and three quarters of seed, but the summer was uncommonly dry and more than one fourth of the crop it is computed has been destroyed by vermin, mice and squirrels, but particularly the latter which are astonishingly numerous and so destructive that scarcely any part of the wheat could be saved. They carried off every ear as it ripened and even before they were ripe. Great numbers of these little animals are destroyed every summer by the boys of the house yet they seem to be as numerous as ever."[37]

Late spring and summer frosts caused serious trouble at all posts. At White Earth House on August 26, 1810: "We have had frost every night since the 23rd. The thermometer at noon has generally stood at 60 degrees. The frost last night froze our potato-tops, and this morning's sun has levelled them to the ground. Mosquitoes begin to be less troublesome."[38] At Edmonton on November 22, 1815: "We have only one hundred & twenty bushels of potatoes from twelve bushels of seed, owing to a severe frost which happened on the 10th of August and entirely destroyed all of the white kind, which made only one half of the quantity planted."[39] Again on June 21, 1828: "Valle with the women weeding the potatoes, they are all frozen black as charcoal, this will injure and even destroy the most of them."[40]

Insects also were in evidence for we note at Edmonton on June 12, 1832: "Insects are destroying our vegetables."[41] And on June 15 the following year: "There are immense numbers of grubs this season which are very destructive amongst the vegetables in the garden they are even commencing their ravages upon the potatoes which now begin to make their appearance."[42] Farther north there also were problems for at Dunvegan, on June 12, 1835: "Innumerable Caterpillars destroy all our Garden stuff Potatoes & Barley as it appears above the ground."[43] And at Fort Vermilion on August 15, 1840: "The infernal Grass-hoppers are destroying my Cabbages, except our Potatoes across the River which they have not yet touched, but in all probability will also destroy them, this is really vexing after all our trouble & care."[44]

Plant diseases also caused problems at a relatively early date. At Edmonton, on September 25, 1823 it was noted: "The Gardens at this place have been very productive last Summer, Barley & Wheat excepted, tho every attention has been paid to both these Crops they entirely failed, the reason of which we cannot with any degree of Certainty assign unless they have been destroyed while in Ear by Blights to which this place is subject even in the middle of the Summer."[45]

Despite all of the limitations the Edmonton post achieved an export position for agricultural products quite early. The first record of this was at Edmonton on April 22, 1820. "At 8 a.m. P.Cunningham & Jas. Forbes set out for Isle a la Crosse in a boat with potatoes, Barley, Horses, and pemican which is to be conveyed across Land from Dog Rump Creek to the Beaver River and from then to their place of destination."[46]

Introduction of Livestock

The time of arrival of domestic animals cannot be determined with accuracy. Horses came into the province before the arrival of the Europeans and Indian reports indicate this to have been about the third decade of the 18th century. Part of the problem in pinpointing the arrival of other stock is that the post journals for some years are missing. Furthermore, there is no assurance that the arrival was actually noted in the journals. It is certain that cattle arrived no later than 1831 for on June 5, 1832 the Edmonton post journal recorded: "Mongrain attending on cattle,"[47] and on July 9: "This afternoon a Cow had a female calf."[48] Because of transportation schedules it is safe to assume that these cattle were at the post the year before and possibly even earlier. Even at this early date consideration was being given to the quality of the cattle for on October 22: "The Bull was this day castrated, as being too small for the increase of our cattle."[49]

Maintaining the cattle was not without its problems for on November 30, 1832: "Old Boulard now has charge of the Cattle, in order to have an eye that the dogs do not attack the Claves (sic)."[50] The cattle were herded in the open even at this late date, for on December 3: "Boulard attending the cattle on the flat point above the Fort by day, and at sunset brings them home so as to get them indoors for the night. This morning a Cow brought forth a female Calf."[51] That this attention was warranted becomes evident when it was noted: "The dogs of the place attacked the cattle and killed a fine young heifer."[52] This was not the end for the next day: "The dogs continue to make war on the cattle whilst we in return wage war against them — two of the rascals were sent to Davie Jones' locker when attempting to kill another cow today."[53] It was not only at Edmonton that dogs were a problem for at a later date, at Fort Vermilion, it was noted: "Put blocks of wood on the dogs necks as Lisotto intended to put out the year-old calves tomorrow."[54]

This practice apparently was not effective restraint for: "The dogs drove one of the year-old calves into the river & unfortunately, before we could render any assistance, the poor little beast was drowned. Blocks of wood round the neck are not sufficient to keep the dogs quiet — a good yard must be put up for them if we are to raise cattle here."[55]

There must have been a fairly substantial herd at Edmonton by 1833 for on May 28 of that year: "Loger and two half breeds from Lesser Slave Lake took their departure. They have with them three cows and a heifer which are to be sent to Athabaska."[56] Again in 1835, in the Dunvegan journal under date of August 23: "B. Bossom & son, assisted by a half breed & a little Boy from L. Slave Lake arrived with cattle from Saskatchewan namely, 3 cows, 3 heifers & 2 Bulls."[57] Some of these were destined for further travel for the following day it was noted: "B. Bossom & our men making a raft to take down some of the cattle (1 cow, 1 heifer & 2 Bulls) to Fort Chipewayan."[58] Apparently the herd at Dunvegan continued to grow for on October 7, 1839, after a fire had destroyed some of their hay, it was recorded: "Sett the men to different duties about the Establishment particularly in making Hay for our Cattle having 17 head to feed during the winter."[59] It is evident that these were kept not only for beef for on May 5, 1840, in a shipment of material to Fort Chipewyan, the cargo included "2 large & 1 small Kegs Butter — 20 Kegs Potatoes."[60]

The first record of pigs at the Edmonton post was in 1826. The first reference is indirect, on October 26: "our hunters supply their families with fresh meat and we give them potatoes of which we gathered 1300 Kegs more than we could destroy among ourselves in any shape whatever, our Pigs require a great many to keep them for the winter."[61] More specifically, it was noted, December 1: "Vandally making a Stye for the pigs."[62]

The horse populations at some of the posts were quite large. Most of the horses were wintered out in the open but a few were kept at the posts. It is evident that serious problems were encountered for on April 9, 1825 the *Edmonton Post Journal* reported: "The Company lost about 100 horses this winter of which one of them died this day of a sickness that there is among them."[63] What the disease was is unknown and whether it was the same as reported the following year is open to question. That year it was recorded: "Arrived from Fort Assiniboine with 23 horses very poor they have been attacked with a disease in their feet which deprived several of them their hoofs and consequently caused death."[64]

There is no record of sheep in the early years in the north and it may well be that this class of stock first entered the province in the south. McInnes states: "a few sheep were driven into the country in the late seventies. Joe MacFarlane for example, placed five on the pioneer ranch, but before the eighties no sheep-ranching industry was established."[65]

Agricultural activity continued to expand at the various trading posts; at the Edmonton post, by 1833, it had become quite substantial. The journal of April 29 and 30 of that year records the following: April 29 — and all the other men with the women were planting potatoes of which 164 Kegs were put in the earth. The covering of these is with wooden hoes and rakes, by the men. April 30 — The same employment for all hands. Today the remainder of the seed potatoes were put in the ground making in all 236 Kegs of cut potatoes, besides which twenty Kegs of Wheat & Barley are also sowed since ploughing commenced and the whole of the cultivated ground is about thirty five acres."[66]

The first agricultural development not directly related to the fur trade occurred at the Lac Ste. Anne Mission established by Father Thibault in 1843 or 1844. As reported by MacGregor: "Here, in a soil that could be cultivated and near the Hudson's Bay Company's old fishing station, he hoped to induce the Metis to settle down to farm."[67] Another mission was established later and of this the *Nor Wester*, printing news from the interior in 1867, reported: "In the fields of the Mission of St. Albert near Edmonton, the crops gave a good result: 300 bushels of wheat, 100 of barley, and 700 of potatoes. In another place, Lac La Biche Mission, besides 200 bushels of wheat, 80 of barley, the Missionaries gathered 720 bushels of potatoes, the produce of 32 bushels they planted. Besides this, a large supply of vegetables were laid in, in both places."[68]

McDougall, at the mission at Victoria, wrote of 1864: "Father, and those who remained at home with him, are endeavoring to teach agriculture as one of the lessons of Christianity. Some seed has been hauled by dog-train from Lac La Biche, and from White Fish Lake in the north, also from Edmonton in the west. A few garden seeds have been carefully put away by the thoughtful mother. A small portion of turnip seed is doled out by the thimbles full. All the hoes the mission party can scrape up, and the one plough they possess are constantly worked, and the beginnings of the mission farm and the first garden patches of the Indians are the result."[69]

At this time all activity of this nature was confined to the northern part of the province. As McDougall stated, in 1863: "The whole country south and west of Edmonton was entirely devoid of settlement; not a solitary settler could you find in all that region. There was not even a trading post south of the Saskatchewan river."[70]

A slight change in this situation was noted the following year: "The woodwork of a plough, made out of birch, was put together at Victoria. This was taken to Edmonton, and there as a great favor it was ironed by the Hudson Bay Company's blacksmith, then taken on Dog-train from Edmonton to Pigeon Lake."[71]

McGregor suggests that the first commercial farm in Alberta was that of Geo. Gunn and George Flett, in 1863. Flett, in writing to the *Nor Wester* had stated that he and Gunn had left the mines to farm. "We seeded 9 bushels of wheat and barley."[72] Not really a big beginning.

Agricultural development continued in the north, accompanied by the introduction of improved equipment. Thus it was recorded, in the *Saskatchewan Herald* of August 25, 1878: "An eight-horse power threshing machine consigned to Edmonton passed westward lately."[73] On September 9 it reported that the crops at Edmonton had turned out magnificently and followed this up on October 7 with the report that: "there are thirty thousand bushels of grain around Edmonton," with two threshing machines reported in operation.[74]

The next year it was reported that five self binding reapers had been ordered for farmers in the Edmonton area.[75] The *Edmonton Bulletin* of December 20, 1880 provides a good picture of the developments relative to farm equipment. "Agricultural machinery was almost unknown until three years since when a combined reaper and mower and one mower did all the cutting that was required. The scythe and the cradle were the levelling implements followed by the time honored flail or thrashed with horses."

"Some idea of the change in the above state of affairs can be had by consulting the following statement of the amount of farm machinery imported since the first of May last: 1 steam saw, shingle and threshing machine, 1 ten horse power threshing machine, 1 self binding harvester, 11 reaping machines, 10 mowing machines, 10 sulky hay rakes, 73 ploughs, 11 fanning mills and 19 iron and wooden harrows. This represents a value of nearly $16,000. In addition to this the Hudson Bay Company has erected and will shortly have in running order a 50 horse power steam saw and grist mill, while messrs. McLeod, Norris & Belcher have erected one of 25 horse power."

"All the above machinery is the best on the market, quality not cheapness being the object aimed at by the purchasers. This is a good exhibit for a new settlement struggling as it is against high freight rates and one thousand miles of bad roads between it and civilization. Although our progress during the past year has been great, it will be much greater in the year coming. Claims are being taken up every day by men who intend to farm them. Goods of all kinds meet with a ready sale at good profit, and every branch of business, notwithstanding a bad harvest, gives promise of continued prosperity."[76]

All of the equipment had to be imported and to ensure that the farmers of the area would be aware of their products, the A. Harris & Son Company, manufacturers of farm machinery at Brantford, began advertising in the *Bulletin* in early 1881 (the paper started in December 1880). A full line of equipment was listed, including "self binding harvesters for both cord and wire."[77]

By current standards, development of farming operations was still quite small, but in terms of the situation at that time some of the operations were fairly substantial. It is not possible to get a complete and clear cut picture of the scope of the operations but some indication is possible. One of the larger operators was Wm. Cust, who not only farmed, but was a dealer as well. On January 3, 1881, it was reported that he had made arrangements to have 5,000 bushels of wheat ground at the new mill.[78] (He had a contract to supply flour on government account to the Indians.) At the same time he was reported to have 200 acres of land fall ploughed. The statistics for 1881 suggest that he very likely was one of the larger operators for there were 267 occupiers of land who had 13,985 acres of land but only 7,615 acres in field crops that year.[79] This suggests an average of about 30 acres of seeded land per operator.

The farmers in the area were not only progressive in terms of the new equipment obtained, but they also were concerned with improvement in other respects. One step in this direction was the formation of an agricultural society in 1879, which held the first exhibition in Edmonton on October 15 "at which over $900 were distributed in prizes, leaving a cash balance in the hands of the society of $160 which has been expended for seed grain."[80] In 1881, imported seed was available for distribution: "Of garden seed there are 34 pounds, including all common varieties. Also half a pound of apple seed, three pounds of White Dutch clover, and twenty small bags, about fifty pounds of spring wheat. Five of the bags are Arnautka wheat, five white Fife, and ten White Russian or Lost Nation Wheat."[81]

In 1884, the annual meeting of the Agricultural Society gave attention to the need for more improved wheat seed, with emphasis on good Fyfe. The reason for concern at this time was that "the time has come when native flour will have to compete against the imported article and for this to be done successfully the best wheat is necessary."[82] Arrangements were made to bring in forty bushels of seed from Winnipeg. The delivered price was $3.60 against a purchase price of $1.20 at Winnipeg.[83]

In the meantime a Fort Saskatchewan farmer had tested the adaptability of winter wheat and G. Rath received the first importation of timothy seed in 1881.[84] Even prior to this some of the more venturesome had imported "eighteen grape vine and seven rose-bush cuttings" to be tested by Messrs. Ross, Hardisty and Reid.[85]

Native Hostility Delayed Southern Development

It is pertinent now to turn attention to developments in the southern part of the province. There was no settlement of any kind south of the Saskatchewan River prior to 1864. Some attempts had been made to establish trading posts, but the enmity of the Indians had made these unprofitable or untenable and they had been abandoned. There is evidence of transient passage of white people through the area, including prospectors going to the gold fields of the north. Father Lacombe pioneered the route from Edmonton to Fort Benton in Montana in 1869, passing through the hostile Indian territory. At about the same time American adventurers and traders were making their way north from Fort Benton into southern Alberta.

Most authorities date the beginning of agriculture in southern Alberta from the establishment of Fort Whoop-Up on the Oldman River in 1869 though it is not stated that any agricultural activity started that year. The first record of agricultural activity at this post was in 1874 when the Mounted Police made their first visit to the fort on their arrival

The Trail Blazers of Canadian Agriculture

in the area: "Mr. Davis, who was by this time in charge, received the policemen cordially and did his best to make their visit as pleasant as possible. He showed the officers over the place, and gave them an excellent dinner, with fresh vegetables from his own garden."[86] This suggests the early 1870s as the beginning of agriculture in the south, nearly 100 years after the beginning in the north.

The first major move came from the north when the McDougalls brought a small herd of cattle to the new mission at Morleyville in 1873.[87] The following year they brought 100 head of cattle from Montana.[88] That same fall (1874) a Mr. Shaw drove about 500 head of cattle across the mountains from British Columbia into Alberta heading for Edmonton.[89] He spent the winter at Morley and apparently stayed in the Bow River country in 1875 for that fall he offered to sell his stock at $38 per head, 60 spring calves thrown in. The stock consisted of 200 cows, including 1-, 2- and 3-year olds; 187 steers, including 1-,2- and 3-year olds; and 9 stout horses.[90]

By this time the police had arrived, bringing with them cattle and farming implements. With their coming the south country began to open up and ranching and farming activities developed at a fairly rapid pace. When the McDougalls arrived from Montana with their 100 head of cattle to augment the small herd brought from the north, they could well claim to be the ranching pioneers. E. Lynch-Staunton makes a different claim when he writes: "Captain Jack Stewart, formerly of the Princess Louise Dragoons of Ottawa and one of the original members of the North West Mounted Police in 1874, with Jim Christie (also an ex-member of the NWMP) founded in 1878 the Stewart Ranch on Pincher Creek, and this undoubtedly was the first ranch to be established in Alberta."[91]

On the other hand we find: "The first large herd of cattle came to High River in the fall of 1882, but around 1876, Lafayette French and O.H. Smith established themselves immediately west of what is now the Town of High River. This location was on the old Fort Benton Trail and as well as keeping a stopping place, as they were known in those days, they maintained a small herd of cattle and horses. They broke and cultivated about 40 acres of land. Finer crops than they raised in '83 I have never seen since that time so that they were pioneers in farming as well as stock raising."[92] To this was added: "Buck Smith had, in the 70s, also located immediately west of French and Smith with a small herd of cattle."[93]

Earlier than any of this activity was the movement of dairy cattle into the south. McInnes states: "In 1875, when it was generally known that the police had established law and order, a herd of dairy cattle was driven over the frontier from Montana by Henry Olson and Joe McFarlane, and established a few miles below McLeod."[94] Kelly adds: "They found a ready market at the police fort for butter and other products, butter itself being indeed 'golden'. Seventy-five cents per pound was the ruling price, and they could not supply the demand. Olson and MacFarland parted company shortly after this, Olson going into the country later included in the Piegan Reserve, and squatting upon a homestead in that unsurveyed district."[95]

Among the early ranchers was Fred Wachter, one of the American fur traders at Stand Off prior to the coming of the police. He stayed on and started ranching at Stand Off in 1876 and apparently became quite successful as he was listed among the ranchers of 1881.[96] It was reported that year that he had between 500 and 600 acres under fence and over 100 acres under cultivation.[97]

In 1882 the first regular organized round up took place. The extent of the area taken up by ranchers at that time can be judged by the fact that the round up started at Kipp, near the present city of Lethbridge, and worked west to the foothills.[98] A report in the *Gazette* that fall noted that a 10 horsepower thresher had been brought in and had threshed 6,833 bushels of grain for nine ranchers with four more to be heard from.[99]

While the preliminary developments were taking place in the area centring on Macleod, there were stirrings in the Calgary area, in addition to McDougall's efforts at Morleyville. When the police arrived at the Calgary site to establish a post in 1875 there were two white settlers on the Elbow River. One of these was Sam Livingston and the other a missionary, Father Doucet. Just to the south of this was John Glenn. He and Livingston were the first to grow farm and garden products in the Calgary district. These were small scale operations but were soon joined by others both large and small. One of the first was that of George Emerson who brought cattle to the area in 1876, and commenced a dairy business with the police as his customers.[100]

One of the first large operations was the Cochrane ranch, established on the Bow River west of Calgary. An idea of the scope of this operation can be seen from two items in the *Saskatchewan Herald* of 1881. The first, on May 23, stated: "A large consignment of thoroughbred cattle of the Shorthorns, Polled Angus, and Herefords, arrived at Halifax on the 1st of April. They were selected in England and Scotland by Senator Cochrane and Simon Beatty for the Farmer's Stock Farm, at Compton, Quebec. Besides these the same steamer brought sixty purebred bulls for the Cochrane Ranche Company, whose lands are situated in the Bow River district, this side of the Rocky Mountains, and seventy-five Oxford-Down sheep for Senator Cochrane."[101]

The second item, on October 3, reported: "Twenty-eight hundred head of Montana cattle for the Cochrane Ranche Co. arrived at Bow River today (Sept.3) and the shipment of imported bulls will be on the ranche today; another band of two thousand head is looked for in a fortnight, and another lot of two thousand five hundred will be delivered by the end of the month."[102]

Of significance in these reports is the fact that purebred cattle were imported to supplement the range cattle being brought in from Montana. This interest in purebred cattle was relatively widespread, as there were numerous reports in the papers of the time regarding the introduction of such cattle to the area. For example, the *Edmonton Bulletin* reported, on November 19, 1881, that: "Mr. Geddes arrived at Bow River from Winnipeg with his band of 200 thoroughbred cows, imported from Ontario. They travelled via Qu'Appelle, Cypress, and Macleod."[103] This meant that they had been trailed the whole distance from Winnipeg, a not inconsiderable feat. A year later the *Macleod Gazette* reported that: "Mr. Lyndon, who came in with 100 thoroughbred cattle a short time ago, has settled near Leavings, and will continue in the stock business."[104] And again: "The North-West Cattle Co. brought in 3,000 head of cattle for the ranch at High River, including 70 pedigreed cows and 10 thoroughbred bulls."[105]

Of special interest, because of recent developments in crossbreeding of beef cattle, is an item from Lethbridge, in 1886. The reporter had visited a ranch north of the city and stated: "Among the herd we noticed some fine specimens of thoroughbred Durham bulls, also a thoroughbred Holstein bull — valued at $600 — from which animal, the owner informed us, a first-class cross was made with his other cattle."[106] Another item, from Medicine Hat, reported: "Mr. Sandy Morrison arrived on Tuesday morning with two train loads of cattle, comprising 700 heifers and 43 pedigreed bulls, Durham, Hereford and Holstein."[107]

Once the south country opened up, development moved at a rapid pace with an early emphasis on ranching rather than on farming. The basis for this was the open grasslands of the foothills and western plains country where all of this development took place. It was fostered by the government which made available 21-year leases of range land to a maximum of 100,000 acres. The annual charge was $10 per 1,000 acres, later raised to $20 per 1,000. The regulations rated the carrying capacity at one animal for each ten acres but this was changed to one animal for each 20 acres.

The rate of increase in the cattle population is indicated in a report from Macleod, in 1882: "It is expected that there will be 80,000 head of cattle brought in this year."[108] In 1883 the development of the south was summarized in an item from Macleod as follows: "Pincher Creek is the largest settlement in the Macleod country, there being forty ranchers there. Almost all both farm and raise stock. Many ranches have commenced on Willow Creek lately so that it is now settled from Macleod to the Leavings, about 40 miles. The Old Man is settled on the Piegan reserve, twelve miles above, to the lime kiln bottom, ten miles below, Macleod. There are several ranches at Kipp, at the junction of the Old Man and Belly rivers, and several have lately taken land near the coal banks, expecting a town to commence there. There are some stock men on the Kootenay and others on the north side of the Belly. Two houses have been built at Mosquito Creek and High River. Sheep, Pine and Fish creeks are all well settled, at least near the Calgary trail. Most of the ranchers raise both grain and stock."[109]

Another picture of the situation is provided by Blue: "In 1881 Dr. McEachren, the veterinary inspector for Canada, reported that over 30,000 Montana cattle had been imported into Alberta and placed on the ranges in the Bow River valley and the Macleod district. The first general round-up was held in 1881 with W.F. Parker of Macleod as Captain. By 1884 the ranching industry was fully established in Southern Alberta located mostly in the foothill country south of Calgary. Forty-one companies were engaged in the business holding under lease an area of 2,782,000 acres. The North-West Cattle Association petitioned the government to prohibit sheep from running on the cattle ranges. Accordingly an Order-in-Council was passed in October, 1884, defining the territory allotted exclusively to cattlemen."[110] This area included the territory south of High River and Bow River to the eastern boundary of the provisional district of Alberta.

"In 1886 the officers of the Department of the Interior estimated that there were 104,000 cattle on the leased lands of Alberta, besides 11,000 more owned by non-leaseholders."[111] This estimate is high compared to that of the Department of Agriculture showing 90,000 cattle in Alberta in 1887.[112] In any event, there was a remarkable rate of development in view of the fact that the railroad had not reached this area until 1883. Prior to that most of the contact with the outside world had been by way of Fort Benton in Montana.

The early market for cattle consisted primarily of the government, which provided supplies to the Indians, and the construction crews building the Pacific railway. The first export of cattle to Great Britain occurred in 1887 via Montreal. The next year 5,000 head were shipped from the Calgary district. These shipments continued for many years, reaching a volume of 12,000 head in 1896.[113]

Though cattle ranching held the spotlight, horse and sheep ranching also developed during this same period. The horsemen had two markets in view, namely cavalry remounts and work horses for the new farmers coming in as settlers. Like the cattlemen they depended on purebred stock for improvement of the native stock. For example, the *Lethbridge News* of February 12, 1886, in an item from Calgary reported: "Mr. Chipman, of the Halifax Ranche Co., has just completed a large purchase of horses from W.H. Dunham of Chicago. The purchase includes a carload of Percheron stallions. Mr. Chipman has also contracted for some 50 brood mares and geldings and is negotiating for 100 mares to be delivered in June."[114]

The sheepmen also sought improvement as shown by an 1888 item from London: "The Canadian Coal & Colonization Company is sending to Alberta a shipment of 200 fine rams to be mated with ewes being purchased in Montana. A number of pedigree animals are included in the shipment."[115]

An interesting sidelight on some of the early ranchers is that they did not limit their attention to livestock. Thus, "Mr. W.S. Lee is thoroughly testing the fruit-growing qualities of this section. He had two ripe cherries from trees set out this spring and is convinced that fruit will thrive in this country."[116] That was in 1882 and in 1883 the following: "Jonas Jones has received a selection of fruit trees, which, notwithstanding the 'non-agricultural' quality of the soil, he will plant on his lease."[117] This reference to the non-agricultural quality of the soil was very likely added by the editor for he, in editorials, had been arguing against the prevailing idea that the area was not fit for crop production.

Unlike the situation in other parts of the west, where most of the settlement was by individuals with limited means, much of the early ranch development was large-scale, requiring substantial financial backing. The rapid rate of development may be considered surprising in view of the limited knowledge available about the area. However, it was the logical extension of developments south of the border where cattle ranching had been pushing north year after year and finally spilled over into southern Alberta. The arrival of the Mounted Police and the subsequent control of the Indians, coincided with this northern extension of ranching into Montana and Alberta.

The open grasslands of the foothills and plains of southern Alberta attracted prospective ranchers. Warm Chinook winds of winter and relatively limited snow cover minimized the requirement for hay for winter feeding.

While major developments were taking place along the western side of the province there also were developments on the eastern side. The railway was making its way from the east, although as in Manitoba and Saskatchewan, the vanguard of settlers preceded the railway. Thus, the *Macleod Gazette* reported, on October 24, 1882, that "there are about forty actual settlers at Medicine Hat, with good comfortable houses, each on quarter sections."[118] Farmer settlers, as distinct from ranchers, were moving into the Macleod area as well.[119]

The coming of the railroad into southern Alberta did not bring with it the rush of settlement that had been experienced in Saskatchewan where the population doubled between 1881 and 1891. In Alberta this decade brought an increase of only 50 per cent. Part of the reason for the slower influx of settlers possibly was the fact that homestead land still was readily available in Manitoba and Saskatchewan and settlers saw no need for going farther afield. Settlement tended to follow railways and it was not until a line was extended from Calgary to Edmonton, into the more attractive farming country, that major increases began. The lack of land survey and any means of officially registering ownership was not as serious a problem in Alberta as in Saskatchewan as the survey had been completed before the major influx of settlers occurred.

Most of the settlers came as individuals but settlement by groups of immigrants began in 1887 when the Mormons arrived and settled in the Cardston district.[120] A group of Germans settled near Medicine Hat in 1889, but they soon became disillusioned with that area and moved to a new location east of Edmonton in 1891.[121] A group of Icelanders settled on the Red Deer River some ten miles west of the Calgary-Edmonton trail in 1889[122] and a group of French Canadians came to settle in the St. Albert district north of Edmonton.[123] They were followed by groups from Ontario who also settled east of Edmonton. According to Blue, these were the years when immigration to Alberta really began. "In 1892, 3,134 settlers took up land along the Calgary and Edmonton Railway, 984 being from Eastern Canada, 620 from the United States, and 220 from the British Isles."[124]

Irrigation an Important Factor

From the very early days of settlement in southern Alberta, irrigation has been an important factor in development, beginning with small-scale, private systems and developing into the major systems that exist today. Authors differ on who the irrigation pioneer was. MacGregor claims that John Quirk irrigated several hundred acres on his Sheep Creek ranch in 1878 while John Glenn did the same on his ranch at Fish Creek in 1879.[125] Blue, on the other hand, gives Glenn the credit with a more modest 15 to 20 acres about 1878, whereas he lists Quirk as a latecomer in 1891. He credits two Americans with a development on the Piegan reserve between these dates and mentions a development, in 1899, on the north slope of the Cypress Hills.[126] The Mormons started a small-scale system in 1890.[127] The *Macleod Gazette* reported, in 1883, that several ranchers at High River had combined to run a ditch through their ranches and another intended to put one through a place on the Belly River.[128]

The Macleod Irrigation Company was the first to receive a charter by act of the Parliament of Canada, in 1891. In 1892 the High River and Sheep Creek Irrigation Company was incorporated and the Alberta Railway and Coal Company received authority to construct irrigation works under its charter. In 1893 charters were granted to the Alberta Irrigation Company, the Calgary Hydraulic Company, and the Calgary Irrigation Company.[129] All of this took place before the Northwest Irrigation Act was passed in 1894 to control the right to use water in the Territory. At the end of 1895 there were 112 ditches with the capacity to irrigate 75,270 acres. Three years later this had increased to 177 and 103,404, respectively. The primary purpose of the first irrigators was to ensure an adequate supply of hay for wintering livestock, but later irrigation was applied to other crop production as well.

The early settlers in Alberta encountered problems similar to those of Manitoba and Saskatchewan; however, ranching brought its own problems. While the climate in southern Alberta was generally favorable to this enterprise, there were some winters when heavy losses were experienced. The pattern can be seen in periodic reports. From Bow River, on April 4, 1881: "There has been little or no loss of stock this winter, and the cattle are as fat as in the fall."[130] On April 29, 1882, again a favorable report: "Cattle as a rule wintered well, but some lately arrived herds have suffered considerable losses."[131] But the next year told a different story: "Considerable hay was put up in the vicinity of Calgary last summer, but when hard weather set in people were unable to haul it home and the Cochrane ate the greater part of it. Hay is now scarce and is worth $25 a ton."[132] An editorial lamented the shortage of cattle at Edmonton and ascribed this to the fact that there was no good supply from the Cochrane ranch because of the severe winter.

On February 9, 1887, the *Lethbridge News* reported: "The unusual severity of the winter and the heavy losses which the cattle ranchers are sustaining, threatens to affect the prosperity of the whole of Alberta."[133] The extent of the losses, not so severe as expected, can be gauged from a report on April 6: "Doc Frields, of the Walrond Ranche, passed through here on Friday last on his way to Montana. Doc says he thinks the losses in stock will not foot up more than ten per cent all round."[134] In contrast to reports of severe weather the following item appeared in February 1886: "Mr. John Nelson brought into our office today some live native grasshoppers. He says the hills are just alive with them. This shows what kind of weather we have in this part of the country, and speaks volumes for it as a cattle district."[135]

Livestock diseases occurred at an early date. Reference already has been made to the heavy losses among the Hudson's Bay Company horses in the north, reported in 1825 and 1826. In 1881 an epizootic among horses was reported in the Edmonton area.[136] In 1889 Mollison mentioned mange and influenza among horses and rheumatism among foals and yearling colts.[137]

The sheep population was reported to be affected by scab in 1893.[138]

A disease resembling foot and mouth disease affected cattle around Slave Lake and Peace River in 1881. The disease was reported to be prevalent in British Columbia as well, but fortunately, mortality was low. Mollison referred to 'big jaw' of cattle in the south.[139] The most serious disease appeared in 1888, or possibly even earlier, on the ranches when an epidemic of blackleg occurred, with losses of 20 to 25 per cent being reported from some ranches.[140] This did not receive much publicity until late 1888 or early 1889 for fear that it might lead to a closure of the British market to Canadian cattle just as this market was being developed. It is evident from the newspaper reports that little was known about the disease and it was discussed as being synonymous with anthrax. Opinions differed as to whether it was infectious or contagious or merely the result of over-abundance of feed.

Weird and wonderful control and treatment procedures were outlined in letters to the editor and in editorial comments. Bleeding was a fairly universal suggestion, supplemented with various treatments such as dosing with coal oil or rubbing with lime. Inoculation with the 'poison of the disease', with garlic, or with 'attenuated anthrax', at branding time, were other suggestions. Some of these remedies were taken seriously for a report, on June 13, 1889, stated: "Quite a number of stockmen, including some of the larger companies, are inoculating their calves with garlic as a preventative against blackleg."[141]

The situation was serious enough that a special meeting of the Stockgrowers Association was held at Macleod on August 1, 1889 to hear a paper by Dr. McEachran, the Dominion Veterinarian. Unfortunately, he was unable to provide a solution to the problem as a control method had not yet been developed.[142]

Three other problems concerned the ranchers. The first of these were the prairie fires that not only destroyed the hay crop but, of equal importance, denuded the winter ranges and deprived the cattle of winter feed. The second was cattle and horse stealing which the police were unable to keep under complete control. The third, after the arrival of the railroad, was the occasional killing of cattle by trains.

Swine production was relatively minor and generally confined to the farms but an interesting deviation was reported from Lethbridge: "A hog round-up is the latest novelty in the stock business. Mr. O. S. Main has a large number of hogs running loose on the prairie at present and has several men at work rounding them up."[143] This system of production did not develop.

In crop production insects caused some difficulty but apparently were not of major significance in the early years of settlement. An 1879 report from Edmonton noted that "cut worms are playing 'hide and seek' with all our garden truck."[144] In 1880 there was a report of "the fly" attacking turnips at Edmonton.[145] A July 15, 1886 report from Calgary stated: "Grasshoppers are becoming pretty plentiful these warm days. We hope they will curtail their sojourn as much as possible."[146]

The only crop disease to which reference was found, was smut, a disease that followed the cereal crops, especially wheat, across the country.

As settlement in Alberta lagged behind that in Saskatchewan so the weed problem also lagged in developing in the time period under review. The first reference to a weed problem was noted in the north. In commenting on the crops in 1882, the editor of the *Edmonton Bulletin* noted: "One great drawback is that all the old land is terribly infested with weeds, owing partly to the two wet seasons just past, partly to the naturally rank growth that is in the soil, and partly to the careless system or rather want of system that has prevailed in farming."[147]

Criticism of the system of farming has been characteristic of the early development in all of the provinces and one response has been similar in all cases, namely, the formation of agricultural societies. Reference was made earlier to the organization of such a society at Edmonton in 1879 and the first agricultural exhibition that year. Apparently the society lost its drive soon after for on November 5, 1881, the *Edmonton Bulletin* lamented that the "society appears to have gone decidedly dead."[148] Whether because of the prodding by the paper, or for some other reason, the society was reorganized in 1882 and held an exhibition on October 19 of that year.[149] This was followed by exhibitions in 1883 and 1884 and continuously since that time.

A Live Stock, Agricultural and Industrial Exhibition Association of the Macleod District was formed in 1886 and held its first exhibition that fall.[150] Not only did it hold an exhibition but it got involved in importing improved seed wheat. Calgary got into the picture with a society and exhibition that year. The same year the Territorial Council passed an ordinance incorporating agricultural societies and providing financial assistance under certain specified conditions. With this encouragement agricultural societies were formed at a number of localities in the ensuing years.

In addition to other activities, some of the societies became involved in developing exhibits to be used in advertising Alberta to prospective immigrants. The Calgary society had an exhibit at an industrial exhibition in the east and this drew favorable editorial comment from the *Macleod Gazette* with the proposal that the Macleod Association should also become involved in such activity.[151]

Partly as a result of Agricultural Society activity, partly because of individual initiative, and partly because of action by the head of the new Experimental Farm at Ottawa, considerable testing of crops was undertaken. An example of private initiative was noted by the *Macleod Gazette* in a report on a Mr. McFarland's ranch. In addition to his livestock operation, Mr. McFarland had substantial farming activities: "of potatoes he estimates he will have in the neighborhood of 200,000 lbs. Mr. McFarland has another piece of ground cropped with several different kinds of oats, with which he is experimenting."[152] The *Calgary Tribune*, February 8, 1888, reported: "A number of farmers are this year experimenting on early varieties of seed wheat. Mr. Steele has just ordered some Russian wheat, known as Egyptian wheat, and will give it a fair trial. He is also getting up a quantity of Racehorse oats, which are said to mature in eighty-six days and give a large yield."[153] Considerable testing of wheat was undertaken by private individuals with samples supplied by the Experimental Farm. Tests conducted in 1887 apparently had given contradictory results for on February 23, 1888, the editor of the *Lethbridge News* urged that the tests be repeated because of some favorable results.[154]

Another important development was reported by the *Lethbridge News* on May 21, 1886: "The first regular meeting of the Canadian Northwest Territories Stock Association took place on Tuesday last in the Macleod town hall, all districts except Calgary being represented."[155] A year later the paper reported: "The first annual meeting of the N.W.T. Stock Growers' Association was held at Macleod." At this meeting the name was changed to the Alberta Stock Growers' Association.[156]

The early Alberta newspapers showed an active interest in agriculture as had been the case in the other provinces. There was no dearth of attention to agricultural news generated in the various areas and this occupied a fair amount of space. Editorial comment on various aspects of the local situation was common and some attention was given to feature articles from other sources. One striking feature of the southern papers was the number of advertisements by ranchers, identifying their brands and their geographic location.

Legislation affecting the development of agriculture was introduced early in the Territorial period. The various ordinances of the Territorial Council, having a bearing on agriculture, have been listed in the chapter on Saskatchewan and will not be repeated here.

By 1901 agriculture was well established in Alberta and the pioneering phase was coming to a close, though major settlement and expansion was still to come. Data in Table I show the developments in the 20-year period from 1881, when statistics first became available. Data for 1966 are included in the table to show the progress that has been made in agricultural production. It is evident that Alberta has become a major agricultural province. All phases of production of the earlier years, except horse production, have shown significant growth. In addition, important new phases of production, not shown in the table, have developed. These include sugar beets, canning crops, and corn for silage and grain, grown under irrigation, as well as rapeseed grown in other parts of the province.

Table I
Population and agricultural statistics for Alberta

		1881	1891	1901	1966
Population	No.	18,075	26,593	73,022	1,463,203
Improved land	Acres	13,985	--	471,906	27,276,251
Horses	No	--	33,058	93,001	93,729
Cattle	"	--	149,894	375,686	3,439,734
Sheep	"	--	25,769	80,055	301,397
Swine	"	--	5,238	46,163	1,092,672
Wheat	Bu	50,648	95,339	797,839	153,000,000
Barley	"	24,624	90,299	287,343	115,000,000
Oats	"	33,705	577,814	3,791,259	104,000,000
Rye	"	--	230	17,249	215,000
Flax	"	--	753	683	4,900,000
Potatoes	"	34,923	195,411	570,334	3,000,000
Hay	Tons	4,491	46,922	176,889	2,937,983
Butter	Lbs	--	427,981	1,211,079	37,705,000

Data for the years 1881, 1891 and 1901 are from the *Census of Canada* for the pertinent years.

The first six items in 1966 are from the census for that year and the remaining items are for 1965 from the 1967 *Canada Year Book*.

Chapter X
British Columbia

British Columbia is the only part of Canada where exploration and settlement were not limited to an approach from the east. The first recorded approach to British Columbia, by the English, was that of Drake, in 1579, but real exploration did not come until two centuries later. Agriculture was introduced into all other parts of Canada by the French and the English, but in British Columbia the Spaniards were the first Europeans to practise agriculture. They had occupied Mexico and California for a long time and had explored the northern coast. However, it was not until the English and the Americans began to send trading vessels into the northern coastal waters that Spain made any move to take physical possession of the northern coast. Some preliminary attempts were made in 1788 and 1789 but it was not until 1790 that a settlement was established. "At Nootka, Eliza proceeded to rebuild the fort of San Miguel which he garrisoned with seventy-five Calatonian troops under the command of Don Pedro Alberni. Barracks and supply buildings were erected, also a residence for the Governor, a church and even a hospital. Gardens were laid out and enclosures for cattle. Expecting to remain at Nootka over the years, the Spaniards were concerned to make themselves comfortable."[1]

That there was considerable development by the Spaniards is indicated by Captain George Vancouver in his report of his visit in 1792 to take over the settlement from the Spaniards. "Senor Quadra requested to know who I intended to leave in possession of these territories; and being informed that it would be Mr. Broughton in the Chatham, in whose charge the remaining cargo of the Daedaldus would be deposited, he gave directions that the store-houses should be immediately cleared, and begged I would walk with him round the premises, that I might be the better able to judge how to appropriate the several buildings, which for the most part appeared sufficiently secure and more extensive than the occasions required. A large new oven had been lately built for our service, and had not hitherto been permitted to be used. The houses had been all repaired and the gardeners were busily employed in putting the gardens in order. The poultry, consisting of fowls and turkies, was in excellent condition, and in abundance as were the cattle and swine; of these Senor Quadra said he should take only sufficient quantity for his passage southward, leaving the rest, with a large assortment of garden seeds, for Mr. Broughton. Senors Galians and Valdes added all they had in their power to spare, amongst which were three excellent goats; I had likewise both hogs and goats to leave with him; so that there was a prospect of Mr. Broughton passing the winter, with the assistance of the natural productions of the country, not very uncomfortably."[2]

The Spaniards relinquished their claim to the area and left soon after. The English took possession but were not prepared to maintain a permanent settlement and thus the establishment was abandoned the next year.

In the meantime penetration was coming from the east with the explorations of McKenzie and Fraser. Finally, in 1805, in what became known as New Caledonia, the first permanent settlement in the province was established at McLeod Lake in northern British Columbia by the North West Company. The following year a post, Fort St. James, was established at Stuart Lake, in the same general region. The first agricultural activity on record is that of Harmon, in 1811, at Stuart Lake when he noted, on May 22: "Planted our Potatoes & sewed Barley & Turnips seeds etc..."[3] Gardening may have been carried on prior to this time as it seemed to be general practice to undertake such activity soon after a post was established.

There is no further reference to this activity at Stuart Lake until 1817 when, on October 4, Harmon reported: "We have taken our vegetables out of the ground. In consequence of the very dry summer, they yielded but poorly. There were months, during which not a drop of rain fell. Fruit of all kinds has been uncommonly abundant this season."[4] It is rather remarkable that there should be an abundance of fruit in a very dry year.

An indication of the productivity of the area was provided on September 3, 1818, when he wrote: "A few days since we cut down and threshed our barley. The five quarts, which I sowed on the first of May, yielded as many bushels. One acre of ground, producing in the same proportion that this has done, would yield eighty four bushels. This is sufficient proof that the soil, in many places in this quarter, is favourable to agriculture. It will probably be long, however, before it will exhibit the fruits of cultivation."[5]

In the interval between 1811 and 1817 Harmon had been at the post at Fraser Lake and there, too, he had engaged in agriculture. In his journal of May 10, 1815: "We have surrounded a piece of Ground with Palisades for a Garden in which we have planted a few Potatoes & Sowed Onion, Carrot, Beet, Parsnip, Seeds, as well as a little Barley, and also planted a little Indian Corn, but the latter I do not expect will come to perfection, as the nights are too cool and the Summers too short to admit of it, for there is not a month in the Year but it freezes, yet in the Day it is warm, and we even have a few Days in the course of the Summer of sultry weather. The soil in many places in New Caledonia is tolerable good."[6] Here is the first reference to Indian corn but it is evident that Harmon was well aware that he was not in country adapted to corn production.

The activities at Fort St. James continued and expanded to a certain extent. A series of journal items indicate the scope of the activities and some of the problems that were encountered.[7] Thus, in 1820: "May 9. Began hoeing the Ground for our Potatoes although Still very wet. Put two Kegs in the Ground. May 10. put the remainder of our Potatoes in the Ground, in all five Kegs sown. May 13. Finished hoeing the ground for the Barley but the ground so wet could not sow. In the evening sowed the Barley. May 15. Got a piece of Ground hoed & arranged for sowing Onions, Carrots, Beets, etc. got another piece hoed to sow a few Peas. May 16. Three men digging the ground to sow Turnips, Lentier & myself arranged a piece of ground & sowed more Onion & Carrot Seed. May 28. The last Barley I sowed began to appear."

Some of the problems are evident in the next entries:"June 9. There are a kind of small Black flies that destroy all the Cabbages who were coming on remarkably well. June 13. The rain we had last night has destroyed all or half at least of the Beans we sowed, who had a fine appearance. June 17. Froze a little last night which injured the Tops of the Beans but not the Potatoes who are coming fine. June 18. Froze again last night. July 3. Our Turnips are coming up but so very thin that we will have few & the flies destroying a number of them. August 2. This Evening Rained and Blew so hard that it crushed all the Barley, owing to it being Sown so thick, therefore I expect we will have but a poor crop. September 2. Our Potato Tops are all Froze. September 14. Finished threshing and cleaning out the Barley & find we have only 7 1/2 Kegs which certainly is a poor crop for a Keg that was sown in the Spring."

The following spring he encountered a different problem as noted on April 23: "I got the Potatoes put in the Ground last fall for Seed, taken out, and found them Swimming in Water. A few of them are Spoiled, but we managed to get Eight Kegs....but what I am most Sorry for at finding no turnips to plant for Seed for next year." This item is particularly interesting as it indicates that the plan was to produce their own seed.

In 1823 we note the first reference to a plow and harrow being used in British Columbia and also note the recurrence of certain problems.[8] "May 14. In the Evening the men opened the pit in which the potatoes have been put last fall for Seed. There were originally 10 Kegs but many of them have been froze and there are but Six Kegs of them that are good. May 19. The men began to plough the Garden for our Potatoes but the Plough being made of wood and covered with Tin, it is often out of order and time is required to repair it. Yet the ground will be better tilled than it could be with the Hoe. May 21. Etter cutting Potatoes for seed and Deloge Harrowing the Ground. Of the Six Kegs of Potatoes taken out of the last Pit, while Mr. McDonald was here very nigh two are so much spoiled that they are unfit for Seed. May 22. Etter still cutting Potatoes and Deloge ploughing and harrowing. May 23. We this Day began to plant our Potatoes and put a Keg into the ground. May 24.

Busy putting our Potatoes in the Ground, say six Kegs and a half, but it was too late to cover them properly. May 25. The men finished covering the Potatoes and had the rest of the day to themselves. May 26. The men busy in the garden finished our Potatoes and if the season is any way favourable we ought to have a good Crop, having put eight Kegs and a half in the ground and no pains have been spared to till the ground well."

But they were not yet through with their seeding. "May 27. The men busy ploughing a piece of ground for Barley. May 28. Desloge busy ploughing but as the plough is often out of order he looses much time to repair it. May 29. Desloge busy in the Garden — Sowed a bit better than half a Keg of Barley. May 30. Desloge busy some time carting Dung over the Potatoes. July 20. A very cold Night so that our Potato Tops have a touch of Frost. July 21. Another cold Night, which has not left our Potatoes uninjured." But this was not the end of the trouble for on July 29 "Etter began to give the last Hoeing to the Potatoes, but it will be useless labour for they were again injured by the frost last night and the Ground is as hard as a Rock."

It is clear that some of the men must have been good artisans for they presumably made their own plow and harrow, and even more for the journal reported on August 25: "Deloge wrought at a pair of wheels for a cart."

The final entry to be noted for that year is dated August 28: "Etter cut the Barley as it is now Ripe and the Squirrels have already begun and would destroy the whole of it."

But there were years when some success was noted and 1824 was one of these: "October 11. This day Dag Corn, Bap Beauvier and myself began to take up the Potatoes and took up 9 1/2 Kegs. Oct.12. took up 16 Kegs of Potatoes. Oct.13. took up 17 Kegs of Potatoes. Oct.14. took up the remainder of the Potatoes, say 18 Kegs which forms 55 Kegs, the produce of four put into the ground in the Spring. Oct.15. took up a Keg of Beets & 2 Kegs of Turnips."[9]

The developments in the north were an extension of the developments east of the Rockies and communication up to this time had been with the east. At this time the Hudson's Bay Company had control of the territory that is now the states of Idaho, Oregon, and Washington. Company policy had ordained that there should be substantial agricultural development in order to supply food for the various posts west of the Rockies, and also to provide agricultural products to the Russians in Alaska as part of a deal giving the company trading rights in Russian territory. Agricultural production on a fairly large scale had developed at Fort Vancouver, on the Columbia River, and at Nisqually farther north, involving both crop and live stock production. There was a gradual increase in communication with California as a source of supply of seed stock; later, as the railroads penetrated to the west coast of the United States, this north-south axis became the main supply route for the posts and for the early settlers. The mountains of British Columbia constituted a serious barrier to supply directly from the east until the Canadian Pacific Railway was completed.

The developments at Fort Vancouver and Nisqually provided the springboard from which to launch the new developments into British Columbia. It was from Fort Vancouver that men and material were provided for establishing Fort Langley, on the Fraser River, in 1827. The schooner carrying men and supplies for the new fort began to unload on July 30 as noted in the journal: "The Schooner was brought close to Shore and the Horses Landed by slinging them off to the Bank. The Poor animals appeared to rejoice in their liberation."[10]

During the first year the main effort was directed at erecting the fort and buildings, but early the next year major effort was directed at agricultural activities. Beginning on March 10, 1828, the journal records with regularity that the men were employed at clearing land and preparing it for planting. On May 20 the first planting took place when: "Planted 5 Kegs Potatoes today to begin with — afterward men commenced clearing more land." Work went on apace for on June 14: "We have now 75 Kegs in the ground." Not satisfied with this it was noted that on July 3: "Planted 16 more Kegs of Potatoes today on a kind of swamp which is hardly yet dry — it may come to something but another year the ground will be good." The overall operation was successful for on November 15 "By 12 o'clock have today had our potatoes in — the whole crop yielding 170 Barrels = 2010 Bushels after 91 put in the ground."[11]

The extent of the area under cultivation is indicated by the entry in the journal on April 19, 1829: "of about 15 acres now open 5 of them is low meadow, 5 fine mellow ground fit for the plough & the rest full of strong Stumps & roots fit only for the Hoe for many years to come, and pasture for more than a few Beasts, is out of the question."[12] In addition to potatoes, some wheat, peas, and barley were seeded that year, this being the first record of wheat in British Columbia. The production for the year, recorded on August 22, was noted: "Our little crop has been collected this week & tho the wheat is not threshed it may be estimated thus: Wheat Bush. 25, Peas 20, Barley 10."

Production was not limited to the cereal crops and potatoes for the journal records that in 1830 a hot bed was prepared in which was planted "various kinds of melons, cucumbers and even Pumpkins and Gourds; besides a great variety of cabbages."[13]

Expansion of acreage continued year by year and by 1835 John Work could report:"The ground has never been measured, but is reckoned that about the fort about 30 Acres are enclosed, including what was ploughed & under crop last year, and what is being ploughed and put under crop this fall and next spring. The soil appears excellent and after being broke up a year or two will no doubt yield abundant crops, but owing to the great quantity of fern and other weeds and the toughness of the turf it requires great labour to break it up. Indeed so difficult are these weeds to banish, that some of the ground under crop this year appears as if it had never been ploughed. This year, owing principally to the unusual dryness of the season, a good deal of the crops failed and yielded very indifferently. At the fort 200 bush. potatoes were sowed & at the big plain 80 bus.; at the fort 15 bush. wheat, & at the big plain 10 bush.; at Fort 15 bush. Pease, and at plain 45 bush.; at the fort — bush. Barley sowed, & at plain 8 bushls. Some oats and Indian corn were also sowed but yielded indifferently. Mr. Yale estimates that he will have about 300 bush. pease, 200 bush. wheat, and about 50 bush. barley."[14]

Expansion Throughout the Mainland

Gradually agricultural activity increased throughout the mainland of the province at the various posts. McLean wrote in 1834: "I was appointed to take charge of Stuart's Lake during the summer, with four men to perform the ordinary duties of the establishment — making hay, attending to gardens, etc... A few cattle were introduced in 1830, and we now began to derive some benefit from the produce of the dairy. The gardens (a term applied in this country to any piece of ground under cultivation) in former times yielded potatoes; nothing would now grow save turnips, a few carrots and cabbages were this year raised on a new piece of ground, which added to the luxuries of our table."[15]

Development at another post is indicated by the following, written the same year: "The accounts and despatches for headquarters being finished in the beginning of March, I was ordered to convey them to Fort Alexandria, to the charge of which post I was now appointed. This post is agreeably situated on the banks of Frazer's River, on the outskirts of the great prairies. The surrounding country is beautifully diversified by hill and dale, grove and plain; the soil is rich, yielding abundant successive crops of grain and vegetable, unmanured; but the crops are sometimes destroyed by frost."[16]

A couple of years later McLean was at Fort George and wrote: "Farming on a small scale had been attempted here by my predecessor and the result was such as to induce more extensive operations. I received orders, therefore, to clear land, sow and plant, forthwith. These orders were in part carried into effect in the autumn. Four acres of land were put in a condition to receive seed, and about the same quantity at Fort Alexandria. Seed was ordered from the Columbia, and handmills to grind grain."[17]

All of these developments were relatively small but they provided sustenance for the post personnel. More important, they provided basic information about the potential as well as the limitations of the various areas of the province for agriculture. The first major development on Vancouver Island occurred when Fort Victoria was established in 1843 to become the administrative and supply centre for the west. This new fort was to replace Fort Vancouver, in Oregon, now that the latter was to come under American control with the settlement of the boundary at the 49th parallel.

At Fort Victoria land was cleared for crops, and cattle and horses were brought from Nisqually. Finlayson, in writing about 1844 and later years, noted: "Having thus convinced the Indians that we were both able and determined to hold our own here, we employed all the spare men we could to clear land, and gradually got the Indians employed in this way also, giving him regular pay in goods. By the end of 1847, we had at this place two dairies with 70 milch cows each, regularly milked twice a day, with some of these wild Indians as assistant dairymen, each cow giving 70 lb. Butter for the summer; the butter exported to Sitka. The flat on which the town is now built was cleared so that in that year, we had 300 acres of it under wheat, peas, potatoes, etc., the land then being rich gave as high as 40 bus wheat to the acre."[18]

Some idea of the facilities with which this activity was accomplished is given in the first-hand description by Finlayson: "After the Fort buildings were put up, the next objective was to cultivate the land so as to raise food for the maintenance of the establishment, as after the first year, any application for agricultural produce from headquarters would be ascribed to want of energy on the part of the officer in charge, hence every effort was made to be independent of this source. Wooden ploughs were made, with mould boards made of oak, chopped out with the axe. Harrows were made of the same material with oak trees. Horse traces made from old rope got from the coasting vessels. As a favor we were supplied with a few iron plough shares from the Depot at Ft. Vancouver, and our plough moulds we got lined on the outside with iron hoops taken off the provision casks first supplied us. Our grain was thrashed by horses driven round a ring in the Barn. We manufactured flour from a hand steel mill supplied us from the Depot. In about four years from our arrival here we had over three hundred acres of land under cultivation & besides supplying our own wants, delivered about 5,000 bushels of wheat with some Beef & butter to two Russian vessels which came here for supplies for which we got paid by bank of Ex. on St. Petersburg, besides which was furnished from here; the rest of the Cargo for these two vessels was supplied from Ft. Langley on Frasers River (then also a farming establishment) sent here for shipment."[19]

The new establishments were gradually taking over the responsibility of Fort Vancouver, providing food for the trading posts and supplying the material for filling the contract that the Company had with the Russians in Alaska.

One item of interest was the involvement of Indians in agricultural production. Douglas, in a report to London, in 1840, stated: "I may be permitted to mention here as a matter likely to interest the friends of our native population and all who desire to trace the first dawn and early progress of civilization, that the Cowegins around Fort Langley, influenced by the Council and example of the Fort, are beginning to cultivate the soil, many of them having with great perseverance and industry cleared patches of forest land of sufficient extent to plant each 10 Bushels of Potatoes; the same spirit of enterprise extends, though less generally to the Gulf of Georgia and De Fuca's Straits, where the very novel sight of flourishing fields of potatoes satisfies our Missionary visitors, that the Honble. Company neither oppose, nor feel indifferent to the march of improvement."[20]

It would seem that the cultivation of potatoes had become even more widespread and general than indicated above. For example, in 1835, John Work reported from Fort Simpson on the coast: "Some Queen Charlotte Island Indians have been here some time ago and traded 177 bush. potatoes, which is a great acquisition, as they serve to enable people to be fed salt fish."[21] Other purchases also were made at Fort Simpson for the 1835 post journal makes numerous references to trading for potatoes in amounts ranging from 12 to 302 bushels, the total for the year being in the range of a thousand bushels."[22]

Up to this time all of the agriculture had been an integral part of the fur trade, no non-company personnel being involved in this activity, except the Indians just mentioned. The company was not at all interested in having settlers in its territory and, even after the territory became a colony, did nothing to encourage settlement. In fact, the opposite was more nearly the case. On the other hand, settlers were not particularly anxious to settle in this part of the British territory. There was still plenty of land available to the east and the difficulty of communication with the outside world was a deterrent.

The first independent settler to the province was a Captain W.C. Grant who arrived in June 1849, with a party consisting of eight men.[23] This did not presage a great rush, for when the first Governor of the colony arrived the next year, there were not more than 30 settlers.[24] But a beginning had been made and by 1853, 19,907 acres of land had been applied for. Of this 10,172 had been claimed by the Hudson's Bay Company, and 2,374 by the Puget Sound Agricultural Company. As the latter was a subsidiary of the Hudson's Bay Company it is clear that that company still played a major role in the colony. Only 7,261 acres had been claimed by private individuals.[25] The development on the island was slow but Begg notes that in 1858 the Puget Sound Company had three well stocked farms in the neighborhood of Victoria. "None but the best breed of cattle, horses or sheep were imported, and the machinery used was of the most improved kind. Crops were generally good, but better adapted for stockraising than for grain. Vegetables did remarkably well. At the settlement of Craigflower, about two and a half miles from Victoria, there were from fourteen to twenty families, a well-culti vated central farm with saw mill, oatmeal mill, etc."[26]

Gold Rush Stimulated Agriculture

In a very direct way the gold rush, beginning in 1858, was the stimulus needed for agricultural development. The influx of miners provided new, strong markets for agricultural products as well as stimulating business generally, the latter in turn attracting new immigrants. A continuing deterrent to settlement was the lack of surveyed land and the absence of a land policy that would encourage settlement. This unsatisfactory situation caused much editorial comment from the editors of the first newspapers in the colony. The *British Colonist*, the first paper, had in the first number, on December 11, 1858, an announcement that "a proclamation, dated Dec. 3rd, having the force of law, is stuck up on the Fort gate, giving power to the Governor of British Columbia to convey lands in that colony."[27] Unfortunately, the basis for conveyance had not been clarified and the lack of surveys complicated the situation.[28] If a settler occupied and cleared an area of land there was no assurance that he would be able to retain ownership.[29] Despite the difficulties, public land was offered for sale. For example, in June 1859 the land office offered land in various districts in the vicinity of Victoria. This land was to be sold at auction, on the first of August, with an upset price of one dollar per acre.[30]

Early in 1860 new regulations were promulgated and it became possible for settlers to pre-empt land, that is, to settle on up to 160 acres of unsurveyed and unoccupied land. Gradually, changes were made in land policy and land was surveyed.

Some of the earliest settlers who had acquired land, began to offer it for sale, two farms being advertised in 1859. The first of these was: "Farm of 385 acres of the richest description of arable land, near the mouth of the important harbor of Esquimault. About 60 acres have been broken and fenced in."[31] The second one was somewhat smaller: "Farm for sale — about four miles from town — 110 acres mostly prairie, part of the land under cultivation."[32] There was no information regarding the price at which such farms were being sold.

In the meantime there were developments on the mainland as well. Scholefield states that "so far as can be ascertained the very first application for farming land was made in November 1858, when W. K. Squires applied for 100 acres for agricultural purposes on the island opposite Fort Hope."[33] Squires was followed by a number of others and the *British Columbian*, in its first issue, on February 13, 1861, in referring to the road from New Westminster to Burrard Inlet, stated that it "passes through a beautiful agricultural district, and leads to the farms of Mr. Holmes and Col. Moody, R.E.[34] On April 4 an editorial stated: "There is within ten miles of this city sufficient good agricultural land for the accommodation of at least five hundred families — say

The Trail Blazers of Canadian Agriculture

100,000 acres.... We have now the nucleus of some five or six populous districts within the same number of miles viz. Mr. McLean on Pitt River, Mr. Atkins on the Quoquitlum, Mr. Holmes on the North Road, Mr. Welsh on the Douglas Road, Mr. McKee below the mill and Mr. Kennedy on the south side of the Fraser."[35]

On July 11, in a letter from Lilloet, it was reported that "crops look well; potatoes are in bloom, and Brady say he will bleed his on the 4th for the feast of Independence. There are about 75 acres of potatoes planted about here. Oats, barley, and wheat are in full ear, and promise a good yield. Brady's Indian corn, onions, turnips, cauliflower, and peas are excellent. The town is well supplied with milk, butter, and vegetables at reasonable rates." At the same time a letter from Quesnel stated: "I have taken up a piece of ground on Swamp river for a ranche, which I intend cultivating in conjunction with my mining."[36]

In 1862, a letter from Lilloet stated that the Pemberton valley was settled and had been under cultivation for a year or two. Hay was selling at $100 to $200 per ton and one farmer claimed to have realized $10,000 from his farming operations.[37] On January 3, 1863, an item from Chilliwack reported that there were 60 settlers in that district which had no less than 40,000 acres of excellent land, half of this being rich, dry prairie providing good feed for animals.[38]

All of this indicates that bona fide farming activities were spreading throughout the province. The Okanagan valley, which today is the main fruit growing area of the province, saw its first settler in 1861. Possibly the best general description of the country and of agricultural development to the mid sixties, is provided by Brown, in an essay written in 1863. Some excerpts from this follow. About New Westminster he wrote: "As regards agriculture there are several farms in the neighbourhood. The soil, though not everywhere deep, is generally very fertile. Of its fertility, the luxuriant vegetation is at the same time the cause and effect. The land has been found to bear abundantly whatever has been tried, especially vegetables and fruit, but owing to so great a portion of the district being densely wooded, the portion of available land in this section of the country is at present limited."[39] As for Pemberton: "There is a fine tract of prairie land in the Neighbourhood of Pemberton known as 'the meadows', it is 7 or 8 miles long, and from half a mile to a mile wide. The land is fertile and produces grass abundantly; it is also well suited for cultivation. There are now 12 farms taken up."[40]

Farther north, at Bridge Creek: "Here a farming country begins, superior to anything seen since leaving Langley on the lower Fraser; the soil is good, and there is abundance of water and wood. From Bridge Creek to Williams Lake is fine country, well adapted for farming; it is said that late frosts might sometimes injure the crops, but at Lake La Hache excellent crops of barley and wheat were grown last summer, and at Williams Lake there is an extensive and productive farm."[41] "At

Beaver Lake there is a pretty large extent of land capable of cultivation. Two or three farms are already taken up, and prove productive."[42]

Of the Okanagan he wrote: "About the centre, on the eastern side, is the Roman Catholic Mission in the midst of an extensive farming district. There are here about 10,000 acres of clear land, having excellent soil, adapted for raising stock, or growing corn, or any kind of produce. There are a few small farms taken up in the neighbourhood of the Mission."[43]

He continued: "There are thousands of acres of good prairie land on the lower Fraser well adapted for stock raising, which is the chief thing to be done in farming in this district. A farm below New Westminster comprises 1,500 acres; there cattle fatten rapidly, and whatever is sown grows well. Close by, is an island with many thousand acres of clear land; the whole comprising 25,000 acres. There is also prairie land at Mud Bay, 10 miles S. of New Westminster; at Pitt River 6 miles to the N., at Fort Langley 15 miles up stream, etc."[44]

"In many places the supply of rain is inadequate, and irrigation has to be resorted to; for which purpose streams abound everywhere."[45]

"With regard to fruits, it is worthy of note that melons grow in the open air without manure, attaining great size and fine flavour: tomatoes also come to full maturity when sown not too late. The orchard at Fort Langley is a great success and it cannot be doubted that this is a good country for apples: orchards are in the course of being planted in various localities, which may one day vie with those of California and Oregon."[46]

Brown dealt briefly with livestock production:[47] "In summer the cattle need little care and no feeding — even in winter they have till last year been left to forage for themselves. Yearling calves and foals, not over 6 months old, have weathered the wintry blasts. But to make no provision whatever against severe weather is at once imprudent and inhuman. Much is not required to be done, a log-built shed for shelter, and six weeks' feed, would save all risks. And the settler can easily obtain hay, grass being everywhere abundant. For sheep the country is found to be admirably suited. It is only a year since the experiment has been made, and it has been attended with complete success. The Southdown thrives best: these may be purchased at Victoria, or cheaper in Oregon."

"The small cactus, which some have erroneously supposed would prove an insuperable obstacle to raising sheep, is most serviceable in the fattening of pigs. In fact these animals require no other food in the summer time than the roots, grasses, and fruits which abound in the woods and plains."

To this Macfie added, with reference to pigs: "Fern-roots, which teem on this island, afford staple food for the last-named of these animals. But to keep them tame and prevent them from being lost in the woods, they should have a stated feed of peas once or twice a day."[48]

Brown made an evaluation of the agricultural potential of the province: "The conclusion as regards the agricultural and pastoral capabilities of British Columbia then is this:

(1) as an agricultural country, it can never be great, or ever vie for instance with California or New Zealand. British Columbia is chiefly not an agricultural but a mineral and mountainous country....

(2) As a pastoral country, on the other hand, British Columbia has great capabilities."[49]

Fraser Valley "Best in the World"

Additional insight into the developments that were taking place can be had from a number of newspaper items. Of particular interest was an editorial: "The delta of the Fraser, comprising, we suppose, one or two hundred thousand acres, when dyked, drained and cleared, will probably compare favorably with the very best agricultural land in the world. They will to a great extent require the treatment given to the flat low lands of the Netherlands, around the mouth of the Rhine, the Fens of Lincolnshire, the Cumberland marshes or the Bay of Fundy, or the bottom lands of the Mississippi, but they will, when reclaimed and protected from the river and the sea, be invaluable."[50] In this the editor was prescient as the lower Fraser area became the most productive area in the province.

The relatively rapid development and the relation of the mining industry to agriculture can be seen from several news items. "We were agreeably surprised to be assured by a gentleman belonging to this city who has a very deep interest in the matter, and is in a position to know, that there are now 250 farms under cultivation on the Lower Fraser, and that there are still excellent farms waiting for at least that many more."[51]

"The Chilliwack Settlements — We learn that a considerable breadth of ground will be put under crop this season. Reese & Co. are preparing to seed 40 acres, and Marks & Co. are also making arrangement by which they expect to seed 30 acres and put up a large quantity of hay, for which there is always a good market at Douglas."[52]

"The breadth of land now under crop in the upper country promises an ample supply of such articles as are being raised. The articles of hay, barley, and oats, used in large quantities for the support of from 6,000 to 8,000 pack animals, were formerly imported, and reached the most fabulous prices by the time they arrived at their destination. Now there will be sufficient grown almost on the spot to meet the demand."[53] The fabulous prices referred to included hay at $100 to $200 per ton[54] and barley ranging from 50 cents to 75 cents per pound.[55]

Another item, in 1863: "I observed on my way up, at all the wayside houses, preparations for farming on a large scale, and large tracts of land were being fenced. From Bridge Creek, the southern boundary of the district, to Williams Lake, I have roughly estimated that there will be about 500 acres under crop. Mr. Jeffrey reports some 2500 head of cattle, and about 300 sheep on the Thompson and Bonaparte Rivers, a part of which are now being driven to the mines."[56]

The rate of development in the Sumas area can be seen from the results of 1866 when, in Chilliwack and Sumas 4,850 acres of land had been taken up. Only 653 acres were under cultivation but they had produced 818 tons of hay, 12,770 bushels of grain, 5,200 bushels of turnips, besides tomatoes, melons, and corn in abundance. There were 744 head of cattle in the district.[57]

The gold mining industry was the stimulus to agricultural production. The thousands of miners and transport animals required to keep them supplied provided a ready market for food and feed products. The overall result was that, with this early stimulus, population increase and agricultural development were more rapid in British Columbia than in Alberta during the 1860s, 70s, and 80s.

Macfie provided a general description of the farming methods in use at about this time. "There are open lands in the colony already fit for the plough, and from which a crop may be obtained without the exertion of clearing. But even the richest prairie soil cannot entirely dispense with preparation for ploughing. Where loose surface stones and small boulders happen to be imbedded, they should be first carefully removed. If there be no dense weed or stumps, the land should be broken up, in the first instance, by one or more yoke of oxen, as the farmer may deem necessary. These animals are preferred for strength and steadiness of draught to the ordinary horses of the country."

"If fern prevail on the land, it should be ploughed up in the heat of summer, in order, by the exposure of the roots to the rays of the sun, to destroy them. These with the bulbous weeds, such as crocuses, kamass, etc., should be collected and burned. Fern-land, not required for immediate use, may with advantage be left for hogs to burrow in, as they form valuable pioneers.... After clearing, draining and ditching should receive attention. I am convinced from observation that where the land is level — favouring the collection of surface water — the benefit of good drainage to crops will, in two years, more than make up for the cost."

"Some advise that the rotating of crops in virgin soil should be: after the ground has been left to summer fallow, wheat sown in October: then a crop of peas, oats, or wheat again: and then a fallow made for turnips. By this time it is estimated that the land will be well cleaned. After turnips, a crop of barley or oats should be raised, followed by potatoes. After the land is subjected to this cleaning process, it is advised that it should be manured, and then placed under the four-course system adopted in Great Britain. It may be stated generally, however, that the time for sowing oats, barley, peas, and tares, is from the middle of March to the end of April; and the time for reaping these crops, from the 1st of August to the end of September. Potatoes are planted in March and April, and gathered in early November. Turnips gathered at the same time, are sown in the six weeks between the 1st of June and the middle of July."[58]

Among the forage crops, Macfie listed red, Dutch, Alsike, and crimson clover, lucerne, birdsfoot trefoil, common sainfoin, common tares or vetch, hard fescue grass, sheep's fescue, Italian rye, and common ryegrass.

The fur trade post personnel and the very early settlers were limited to quite primitive farming equipment, much of it home made. On the other hand, more advanced equipment appears to have been on the west coast at quite an early date. McLoughlin, at Fort Vancouver wrote, in 1836: "In the Requisition now sent there is a demand for two reaping machines. In the Encyclopedia of Agriculture there is an account that two of such machines were brought into operation in Fifeshire, and found to be a great saving of labour. I wish you would have the goodness to cause enquiries to be made and if they are as represented let them be sent, they are said to cost only £30; even if they cost £50 each it is no object in comparison with the advantage we would derive from having them to cut down our crops."[59]

He must have been getting some of the equipment that he wanted for the Governor and Committee wrote to him on December 21, 1842: "We cannot help noticing the heavy outlays incurred of late years in the purchase of Agricultural implements — threshing machines, horse tackle etc. etc., which it is desirable to reduce as much as possible; the wood work of Ploughs, we think ought to be prepared in the country, likewise the horse collars, hames and harness; one threshing machine at Vancouver, we should consider quite sufficient for the Company's use."[60]

Though there was advanced equipment at Fort Vancouver, the early reports from Fort Langley and Fort Victoria suggest that they did not have any of it. Even in 1862, some years after settlement began, the only equipment advertised in the newspaper was "patent wrought iron ploughs and Excelsior harrows."[61] A year later a new advertisement appeared in the Victoria paper. A San Francisco company advertised: "Patent Hay presses, mowing and reaping machines, fanning mills, ploughs, harrows, scythes or any other implements or machines."[62] The presence of these machines in British Columbia was yet to come, but on August 13, 1863, the *British Columbian* reported: "Mr. Davidson, of the well known Davidsons Ranch in the upper country, recently went down to San Francisco and purchased a reaping machine, and also a thrashing machine, which he has taken up to his ranch at very great expense."[63]

As time went on more and more equipment became available. For example, an advertisement in 1869, listed: "Threshing machines, Excelsior mower and reaper, New York combined reaper, Union mowers, Excelsior grinder, Price's Petaluma hay press, Pitt's improved threshers, Whitcomb's patent horse hay rake, wire horse rake, revolving wood horse rake, Sweepstake gang plow, as well as various hand tools."[64] A later news item reported that: "Mr. Edgar Mains has been appointed Colonial agent for the celebrated Buckeye mowers and Pitt's threshers. These labor-saving machines are in general use throughout the States and Canada, and our farming friends, if they would compete with foreign produce, must employ the cheapest and most expeditious appliances for harvesting their crops; with hand labour they will always be undersold."[65]

Associations Help Farmers Learn

Meagre as their resources might be, the early settlers were aware of the need for means to improve their practices. As one step in achieving this they followed the practice of settlers to the east, forming agricultural societies and holding exhibitions. The first such society in British Columbia was the Salt Spring Island Agricultural Society formed in May 1860.[66] Apparently it did not hold exhibitions until 1869 when it was reported that: "The settlers of Cowichan, Chemarous and Salt Spring Island held their first annual Fair on Wednesday last at Maple Bay."[67]

However, the first exhibition in British Columbia was held in the fall of 1861 at Victoria with exhibits of crop products. There were no classes for livestock though some were on display. Among these the only breed mentioned was the Devon, and so it may be assumed that breeds, as such, had not yet become important.[68] Other societies were formed as time went on. The Saanich Agricultural Association was formed in 1869 and held its first exhibition that year.[69] Chilliwack had its first society and exhibition in 1874, with Richmond, Surrey, and others coming after that.[70]

The societies not only held exhibitions but undertook other activities as well. For instance, the Saanich Association held its first plowing match in 1870. Part of the report on this match indicates the relative importance of horses and oxen as motive power and provides an interesting insight into the type of prizes offered at that time. "All competitors used horses, except Mr. C. F. Lester who worked with oxen. The ploughing commenced at 10 a.m. and continued until 4 p.m., although several had completed their half acre before 3 o'clock."

The prizes were:[71]

Adult prizes
1st	$10.00	1 hat, 1 set whipple trees
2nd	$ 7.50	1 set whipple trees
3rd	$ 5.00	1 pair team bridles
4th	$ 2.50	1 pair halters
5th	$ 2.50	1 neck yoke
6th	$ 2.50	1 riding bridle
7th	$ 2.50	1 waggon whip
8th	$ 2.50	1 baltic shirt

Youths' prizes
1st	$ 3.00	a riding bridle
2nd	$ 2.00	1 scarf, 1 pair spurs
3rd	$ 1.50	1 pair spurs
4th		2 scarfs
5th		1 scarf, 1 pair gloves

The newspapers of the day paid a fair amount of attention to agriculture and its problems even though gold mining was the major news attraction for many years. The editors editorialized quite frequently on the favorable prospects for agriculture, on the need for government support, and on the value of agricultural societies and exhibitions. They were not above criticising the production practices of the farmers as can be seen from part of an editorial in 1869: "We have said agriculture has long since assumed the rank and dignity of a science. It is problematical how far that is true as applied to British Columbia. It is feared that farming is, for the most part, not carried on in a very scientific way here. Indeed in many instances there is observable a degree of slovenliness and thriftless indifference most painful to behold and most fatal to success."[72] This was an echo of comments general in practically all other pioneer areas of Canada.

The papers also included articles on agricultural subjects though these were not so numerous as in some of the papers east of the mountains.

Live stock production had been a part of the agricultural picture from the very earliest time as the Spaniards had cattle, goats, and poultry at Nootka. However, after they withdrew there is no further reference to domestic animals, except horses, until some 40 years later. It has not been determined when horses first entered the province, and it is not clear whether some of the Indians in the southern interior had horses before the coming of the Europeans. We do know that the fur traders had them in the 1820s and possibly earlier. Their presence at Fort St. James, in 1823, can be inferred from the fact that ploughs were in use that year and there were no oxen in the country at that time.[73]

At Fort Langley, in 1829, ground was prepared for cattle expected that year. It is not clear that they arrived, though in January 1832 McDonald reported that he had four milch cows.[74] The situation changed quite rapidly, for the 1835 journal reported that there were 60 pigs and

20 cattle at the Fort. Of the cattle, five were calves of the year, and six were oxen broken to work. It was reported that the cattle were in good condition but the pigs were rather lean.[75]

Farther north, at Stuart Lake, in 1834 McLean reported that a few cattle were brought there in 1830 and the traders benefitted from their produce.[76] There is nothing to indicate where these cattle came from or how they had been transported, but it may be assumed that they had been trailed from Fort Vancouver or some other post in what is now the United States.

It is known that pigs arrived earlier, for in July 1829, the Langley journal recorded the receipt of goats and pigs.[77] The early experience with these two classes of stock was not successful as can be seen from the journal entries of that year.[78] On October 5: "Our Billy Goat, the father of the flock, was castrated today for mischievous behaviour." Unfortunately this had a fatal effect for a few days later: "Our Billy Goat did not get over the operation."

The pigs also failed to perform satisfactorily. On October 12: "One of our Sows pigged yesterday, but had the misfortune to lose them all — the brute is so fat that the little ones could never get at the milch — however I hope the other one will succeed better, although there is a lot of grease there too." His optimism was not justified for in November he had to report that not one of the 11 young born survived despite all efforts to help them suckle.

A little better success was obtained with one of the female goats as he could report that a young female had given birth to three young and one of these survived.

Poultry were among the first arrivals. It is certain that they were at Fort Langley by 1828, for in August of that year the journal reported the following interesting little item: "Our friend Joe came across with some 20 Beaver Skins he took a great fancy to our Poultry and nothing would suit him but a Couple of them & offered a Hen for cash. However, on the score of good fellowship we gave him a Brace but no sooner had he reached his lodge than the Canis tribe made very free with them & the old Knave sent his son immediately with a dead one to exchange it for an animate one. I promised him one when he leaves the river this fall, and he was well pleased at the idea of its being away from the Dogs till then."[79]

Dogs caused other problems as well, for we learn from the Fort Alexandria journal, June 28, 1827:[80] "As they were arriving with the band of Horses the whole of our Dogs got after the Colts and all the exertions that 4 of us could muster could not prevent them getting hold of one which they would have torn to pieces had we not had recourse to Guns, and fired Several Shots before we could make them leave off. One of the Dogs was Shot dead and 2 others severely wounded." But this was not the only problem with the horses for later it was reported that two of the horses had the disease called staggers.

With the development of Fort Victoria, and the later influx of miners and settlers, livestock became an important part of the agricultural scene. For example, in April, May, and June of 1862, there were imported: Horses & mules 3,434; Oxen 162; Asses 1; Camels 21; Beef cattle 1,139; Sheep 638; Bulls 9; Cows 280; Calves 44; Hogs 26. To this was added an estimated 1,700 sheep to Similkameen. Later that year 500 sheep came in by steamer to Hope and were driven to Kamloops.[81]

According to Vrooman 1862 was the year when ranching began in the Kamloops area with cattle driven up from Oregon along the usual route by way of the Okanagan valley. Ranches were well established along the Thompson River valley by 1865 and had begun to move up the Cariboo. Most of these ranches were relatively small in terms of cattle numbers, the first large ranch being the Douglas Lake ranch established in 1882.[82] This type of development contrasts with southern Alberta where there were relatively large ranches established early in the development of that area.

Apparently most of the early cattle were non-purebreds as the introduction of a purebred bull was of sufficient interest to warrant a newspaper item: "Dr. Tolmie, Chief Factor Hudson Bay Company, has imported from England at a great expense, a very superior Durham Bull, which was brought up by the Steamer Otter on Tuesday last and is to be kept at the Company's farm at Langley. This will afford a fine opportunity to stock-raisers for improving their breeds."[83]

Despite the relatively mild climate of the lower Fraser valley, there were problems with winter livestock losses. The winter of 1861-62 apparently was unusually severe as it was reported that horses, mules and cattle being wintered at Sumas were dying off. It was estimated that half of the animals would die. The reason given for the losses, in addition to the severe winter, was that the animals had been in poor condition going into the winter.[84] In the interior of the province the situation also was grave with losses estimated as high as fifty per cent.[85]

Predators were another cause of losses from to time, an example being in 1870 when it was reported that: "Wolves and panthers, driven down from the mountains by the late severe weather, have been unusually active in the Cowichan district. Sheep have suffered greatly from their depredations and poisoned meat set out to tempt the prowling scamps to a speedy and ignominious end, has victimized the settlers' dogs instead."[86]

The vagaries of weather and other hazards of agricultural production were noted at the various posts almost from the beginning and have been recorded earlier. After settlement commenced, there was practically no reference to insects or diseases affecting crops, a situation quite different from that in other parts of Canada.

Weeds were a problem. In the early stages these were primarily indigenous plants, but later they were introduced from outside. One of the first mentioned was the thistle, though unlike the situation east of the mountains, where the Canada thistle was the culprit, the Scotch thistle was named first in British Columbia. The first official recognition of this was in 1877 when an act was passed to prevent the spread of thistles.[87]

Flooding of land along some of the rivers caused damage from time to time, one of the first floods reported being on the Thompson River in 1870, when "great injury was sustained by the farmers in consequence of the flood; fences, and in some instances young stock being carried away and the prospects of food crops ruined. Beyond the reach of the water the hope of an abundant yield is good."[88]

Fruit production has become a major item in British Columbia and this had its beginning in the very first years of settlement. Macfie makes reference to apple trees loaded with fruit at Fort Langley.[89] The first Victoria exhibition listed apples among the produce shown.[90] An advertisement in 1863, stated: "have on hand a choice assortment of fruit trees etc. consisting of Apples, Pears, Plums, Cherries. Also Gooseberries, Raspberries, Blackberries, and Strawberries."[91] The extent of orchard development can be inferred from the fact that in 1870 one nursery advertised 25,000 fruit trees for sale.[92]

Hops was another crop that was grown and as early as 1870 produced enough for local demand and a small surplus for export.[93] The year before, it was reported that a lively trade in cranberries had developed and all the coopers were busy making barrels to hold them. The production was principally in the lower Fraser valley and the market was in California. One firm had 100 barrels of cranberries for shipment at one time.[94]

Proclamations and Ordinances

Soon after the establishment of colonial government, the governor and the governing council began to deal with matters affecting agriculture and various proclamations and ordinances were enacted from time to time. While two separate colonies were established and existed for some years, the proclamations and ordinances have not been kept separate in the listing that follows. The most significant ones will be listed . Disposal of Crown lands was the most vexing problem in the first few years after the British repossessed the territory from the Hudson's Bay Company.

1858
- Proclamation enabling the Governor to convey crown lands.

The Trail Blazers of Canadian Agriculture

1860
- Pre-emption act. Upset price of land not sold at public auction to be 10 shillings per acre.

1861
- Land registry act.

1862
- An act to prohibit swine and goats from running at large in the town of Victoria, and to prohibit goats from running at large in the settled districts of Vancouver Island.

1865
- An ordinance for regulating the acquisition of land in British Columbia.
- An act to impose landing permit dues on the importation of certain stock and carcasses.

1866
- An Act to exempt the Homestead and other property from forced seizure and sale in certain cases.

1869
- An ordinance to provide for the fencing of land in British Columbia.
- An ordinance for the better protection of cattle and the better prevention of cattle stealing. This included a penalty for fraudulent branding and made provision for registering brands.

1871
- An act to exempt (in certain cases) cattle farmed on shares, and their increase, from the operation of any bankruptcy or insolvency laws.

1872
- An act respecting breeding stock. This restricted stallions, bulls, and boars from running at large in designated areas. It was amended in 1876, to include jackasses and rams.

1873
- An act respecting drainage and dyking, and irrigation of lands in British Columbia.
- An act to incorporate agricultural societies. This made provision for grants proportional to amounts subscribed by members of the societies. This was amended later to require a minimum of 25 members in order to qualify for a grant.

1875
- An act respecting injuries caused by animals of a domestic nature.
- An act respecting the marking of cattle.

1876
- An act to provide for the better protection of cattle ranges. This prohibited sheep on such ranges.

1877
- An act to prevent the destruction of pasturage on the islands in the Gulf of Georgia. This restricted the grazing of sheep to actual owners and also limited the number pastured on crown lands. The maximum was 200 head, 5 months or older, for every 100 acres of land owned.

- An act to prevent the spread of thistles.

1878
- An act for dyking and reclaiming certain lands at Chilliwack, Sumas, and Matsqui.

1879
- An act to protect winter stock ranges.

1881
- An act respecting, the transfer of cattle brands and marks.
- An act respecting dogs. This permitted the destruction of dogs caught in the act of worrying sheep.
- An act to prohibit the owners of swine from permitting same to run at large.

1886
- An act providing for the election of, and defining the duties of water viewers. This was in connection with irrigation disputes.
- An act to prevent the spreading of noxious weeds. This included penalties for importing and selling grain containing seeds of noxious weeds. These were: Canada thistle, oxeye daisy, wild oats, ragweed, charlock, sorrel, burdock, wild mustard, or any other foul seeds. This act also applied to transporting seeds from farm to farm in threshing machines or fanning mills.

Some indication of the support given to agricultural societies and exhibitions is found in the estimates. The item for agricultural societies first appeared in 1880 at $500 and continued at that level for several years. In 1886 it was raised to $750 and in 1888 to $1,000. The item for exhibitions first appeared in 1883 when it was listed at $1,000 for the British Columbia Agricultural Association in aid of the provincial exhibition. This same amount was continued for several years.

One other item of interest was listed in 1883: $500 for destruction of wolves and panthers in settled districts. This was reduced to $100 in 1885 and continued at that level to 1888 when it was increased to $150.

The development of agriculture in British Columbia, prior to the building of the Canadian Pacific Railway, was relatively slow and, for a time, primarily dependent on the miners for a market. Later the building of the railroad provided stimulus and, as industry developed and population increased, the outlet for agricultural products grew. A picture of the development is provided by the data in Table I. This shows a steady but not spectacular growth up to 1891. The data for 1966 are shown to indicate the changes that have taken place. Much of the increase in grain production since 1891 has taken place in the Peace River Block which was not settled at the time.

Brown's conclusion, in 1863, that "as an agricultural country, it can never be great"[95] has been borne out but the agriculture of the province nevertheless fills an important role in providing products for the people of the province as well as fruit for other parts of the country.

Table I
Population and agricultural statistics for British Columbia

		1870	1881	1891	1966
Population	No.	36,247	49,459	98,17	1,873,674
Improved land	Acres	--	184,885	57,881	1,614,141
Horses	No.	4,364	26,122	44,521	26,521
Cattle	"	20,845	80,451	126,919	546,013
Sheep	"	9,064	27,788	49,163	65,919
Swine	"	8,629	16,841	30,764	37,422
Wheat	Bu.	7,518	173,653	388,300	2,600,000
Barley	"	19,373	78,990	79,024	4,500,000
Oats	"	48,671	253,911	943,088	3,000,000
Rye	"	--	482	6,141	51,000
Potatoes	"	31,711	473,831	685,802	2,100,000
Hay	Tons	1,744	43,898	102,146	1,000,000
Butter	Lbs.	--	343,387	393,089	3,303,000

Data for 1870 are from Appendix T, p134, in the *Official Report on a visit to British Columbia* by Langevin, H.L. Minister of Public Works, 1872.

Data from 1881 and 1891 are from the *Census of Canada* for those years.

The first six items for 1966 are from the census for that year. The remaining items are for 1965 from the 1976 *Canada Year Book*.

REFERENCES

List of Acronyms

BCARS   British Columbia Archives and Records Service
NA   National Archives of Canada
PANB   Provincial Archives of New Brunswick
PANS   Provincial Archives of Nova Scotia

Introduction

1. R. Hakluyt, *The principal navigations, voyages, traffiques, discoveries of the English Nation* (Glasgow: J. MacLehose & Sons, 1904) Vol. 8, p. 11.

2. H. A. Innis, *The fur trade in Canada* (Toronto: University of Toronto Press, 1956) p. 136.

3. Hakluyt, Vol. 7 p. 362.

Chapter I — Newfoundland

1. S. E. Morison, *The European discovery of America* (New York: Oxford University Press, 1971) p. 53.

2. D. W. Prowse, *A History of Newfoundland* (London: Eyre & Spottiswood, 1896) p. 59.

3. H. P. Biggar, *Precursors of Jacques Cartier 1497-1534* (Ottawa: Government Printing Bureau, 1911) pp. 196-197.

4. E. Haies, "A narrative of the expedition of Sir Humphrey Gylberte in 1583 for the planting of a colony in America," *Sir Humphrey Gylberte and his enterprise of colonization in America*, ed. John C. Slafter (Boston: Wilson & Son, 1903) p. 135.

5. Lord Birkenhead, *The Story of Newfoundland* (London: Horace Marshall & Son, 1920) p. 67.

6. Prowse, p. 94.

7. Ibid., pp. 126-127.

8. Ibid., p. 98.

9. Ibid., p. 99.

10. Capt. E. Wynne, A letter from Captain Edward Wynne, Governor of the colony at Ferryland, within the Province of Avalon, in Newfoundland, unto the Right Honourable Sir George Calvert, Knight, his Majesty's Principal Secretary, July 1622, NA, 1622-1, pp. 3-4.

11. Ibid., p. 9.

12. Ibid., p. 10.

13. Ibid., p. 8.

14. Birkenhead, p. 77.

15. Prowse, p. 195.

16. A list of inhabitants and their concerns from Trepassey to Cape Bonavista in 1677, NA, Colonial Office papers 195/2, pp. 17-21.

17. Capt. Underdown, Letter of November 25, 1706, to Board of Trade and Plantations, NA, Colonial Office papers 195/4, p. 300.

18. NA, Colonial Office papers 195/5, p. 64.

19. Ibid., p. 235.

20. NA, Colonial Office papers 199/17, p. 68.

21. Ibid., pp. 94-96.

22. Thomas Aaron, *The Newfoundland Journal of Thomas Aaron*, ed. Jean M. Murray (Don Mills: Longman's Canada Ltd., 1968) p. 67.

23. Ibid., p. 74.

24. Ibid., p. 175.

25. *Royal Gazette & Newfoundland Advertiser*, Vol. 6, No. 305, July 1, 1813.

26. Ibid., Vol. 22, No. 1180, April 6, 1830.

27. Ibid., Vol. 25, No. 1341, July 23, 1833.

28. NA, Colonial Office papers 197/4, p. 15.

29. *Royal Gazette*, Vol. 3, No. 144, May 24, 1810.

30. Ibid., Vol. 26, No. 1353, October 15, 1833.

31. Ibid., Vol. 3, No. 152, July 19, 1810.

32. Ibid., Vol. 4, No. 221, November 14, 1811.

33. Ibid., Vol. 20, No. 1078, April 22, 1828.

34. Sir R.H. Bonnycastle, *Newfoundland in 1842* (London: Henry Colburn, 1842) Vol. 2, pp. 7-8.

35. Ibid., pp. 22-23.

36. *Royal Gazette*, Vol. 26, No. 1378, April 8, 1834.

References

37. Bonnycastle, Vol. 1, pp. 169-170.
38. Ibid., p. 171.
39. Ibid., p. 173
40. NA, Colonial Office papers 197/10, p. 12.
41. Ibid., p. 16.
42. Bonnycastle, Vol. 2, pp. 18-19.
43. Ibid., pp. 28-29.
44. Ibid., pp. 34-35.
45. Ibid., p. 37.
46. Rev. P. Tocque, *Newfoundland: as it was and as it is in 1877* (Toronto: John B. Magurn, 1878) pp. 431-432.
47. Ibid., p. 437.
48. NA, Colonial Office papers 197/4, p. 1.
49. Tocque, pp. 430-431.
50. Ibid., p. 437.
51. W. E. Cormack, *A journey across the island of Newfoundland in 1822* (London: Longmans, Green & Co. Ltd., 1928) p. 96.
52. Ibid.
53. NA, Colonial Office papers 199/17, pp. 144-145.
54. NA, Colonial Office papers 199/31, p. 158.
55. Data for 1836, 1845, and 1857 from NA, Colonial Office papers 199; data for 1869 from Abstract of census and return of the population etc. E. D. Shea, printer.
56. Data from NA, Colonial Office papers 199 for the years indicated.
57. NA, Colonial Office papers 199/61, p. 255.
58. Acts of the Newfoundland Assembly from 1833 to 1865.

Chapter II — Nova Scotia

1. R. Brown, *A history of the island of Cape Breton with some account of the discovery and settlement of Canada, Nova Scotia, and Newfoundland* (London: S. Low, Son, and Marston, 1869) p. 18.
2. M. Trudel, *The Beginnings of New France*, trans. P. Claxton (Toronto: McClelland & Stewart, 1973) pp. 63-64.
3. M. Lescarbot, *Nova Francia - A description of Acadia 1606*, trans. P. Erondelle (London: George Routledge & Sons Ltd., 1928) p. 32.

4. Ibid., p. 43.
5. M. Lescarbot, *The history of New France*, trans. & ed. W.L. Grant (Toronto: The Champlain Society, 1907-14) Vol. 2, p. 226.
6. Lescarbot, *Nova Francia*, p. 47.
7. Ibid., p. 90.
8. Ibid., p. 93.
9. Ibid., p. 117.
10. Ibid., p. 136.
11. F. E. Rameau, *Une colonie feodale en Amerique: l'Acadie: 1604-1710* (Paris: E. Plou, Nourritt, 1877) p. 30.
12. Lescarbot, *The History of New France*, Vol. 3, p. 67.
13. F. Parkman, *Pioneers of France in the New World* (Toronto: G. N. Morang, 1899) Vol. 2, p. 139.
14. N. Denys, *Description and natural history of Acadia*, trans. W. F. Ganong (Toronto: The Champlain Society, 1908) pp. 137-138.
15. G. Lanctot, *A history of Canada: From its origins to the Royal Regime 1663*, trans. J. Harnbleton (Clarke, Irwin & Co., 1963) Vol. 1, p. 141.
16. M. Coleman, *The Acadians of Port Royal 1632-1775* (Ottawa: National Historic Sites Service) Manuscript Report No. 10, Vol. 446, p. 1.
17. G. G. Campbell, *A history of Nova Scotia* (Toronto: Ryerson Press, 1948) p. 28.
18. Lanctot, p. 292.
19. Ibid., p. 282.
20. W. I. Morse, *Acadensia Nova 1598-1779: New and unpublished documents and other data relating to Acadia* (London: B. Quarritch, 1935) Vol. 1, pp. 177-178.
21. Denys, p. 177.
22. Ibid., p. 105.
23. Ibid., p. 159.
24. Ibid., pp. 166-167.
25. NA, MG1 G1, Vol. 466.
26. M. Coleman, *Acadian history in the Isthmus of Chignecto* (Ottawa: National Historic Sites Service) Vol. 448, p. 3.
27. Campbell, p. 110.
28. NA, MG1 G1, Vol. 466.

References

29. Ibid.
30. NA, Colonial Office papers 217/2, p. 231.
31. PANS, Vol. 26, pp. 30-33.
32. PANS, Vol. 23, pp. 49-50.
33. PANS, Vol. 7, p. 70.
34. *Halifax Gazette*, No. 4, April 13, 1752.
35. C. Lawrence, "The journals and letters of Col. Charles Lawrence," *PANS Bulletin* (Halifax: 1953) No. 10, p. IV.
36. *Halifax Gazette*, No. 64, June 16, 1753.
37. Lawrence, p. 40.
38. Ibid., p. 29.
39. *Halifax Gazette*, No.114, June 22, 1754.
40. J. C. Webster, *Acadia at the end of the 17th century* (Saint John 1934) New Brunswick Museum, Monographic Series No. 1., p. 128.
41. N. de. Dierville, *Relation of the voyage to Port Royal in Acadia or New France*, trans. C. Webster, ed. J. C. Webster (Toronto: The Champlain Society, 1933) p. 34.
42. Ibid., pp. 108-109.
43. Ibid., p. 109.
44. C. Morris, *A brief survey of Nova Scotia*, NA, MG18 F10.
45. I. Deschamps, PANS, Letter - Deschamps Manuscripts.
46. I. Deschamps, "Sketch of the province of Nova Scotia 1782," in Appendix III, *Report of Public Archives of Nova Scotia*.
47. J. Robinson T. Rispin, *A journey through Nova Scotia*, cited in *Select Documents in Canadian Economic History* by H. A. Innis (Toronto: University of Toronto Press, 1929) Vol 1 pp. 21-34.
48. J. Young, *Letters of Agricola on the principles of vegetation and tillage* (Halifax: Holland & Co., 1822) p. XI-XII.
49. Ibid., p. 162.
50. *Royal Gazette*, Vol. 1, No. 32, November 3, 1789.
51. Ibid., Vol. 3, No. 104, March 15, 1791.
52. J. S. Martell, "Achievements of Agricola and the Agricultural Societies 1818-1825," *PANS Bulletin* (Halifax: 1940) No. 6, p. 6.
53. *Nova Scotia Royal Gazette*, Vol. 6, No. 313, December 9, 1806.
54. *Acadian Recorder*, Vol. 6, No. 46, November. 14, 1818.
55. Ibid., Vol. 7, No. 29, July 17, 1819.

The Trail Blazers of Canadian Agriculture

56. Deschamps Manuscripts 1750-80, PANS, MG1, No. 258, p. 22.
57. Minutes of H.M. Council Annapolis 1735, PANS, p. 318.
58. NA, Colonial Office papers 217 MG11/7, p. 55.
59. King's Printer, The statutes at large passed in the several General Assemblies, Halifax.
60. *Nova Scotia Royal Gazette*, Vol. 2, No. 54, January 14, 1802.
61. King's Printer
62. A. W. H. Eaton, *The history of King's county*. (Salem Press, 1910, reprinted, Belleville: Mika Studios, 1977) pp. 203-204.
63. R. P. Gorham, *Landmarks in early Maritime agriculture*, Mss., PANB, MG04/7/2.
64. *Nova Scotian*, Vol. 7, No. 43, October 29, 1834.
65. Ibid., Vol. 6, No. 41, October 9, 1833.
66. *Royal Gazette*, Vol. 4, No. 174, July 10, 1792.
67. Young, p. 361.

Chapter III — New Brunswick

1. S. Champlain, *Works of Champlain*, ed. H. P. Biggar, (Toronto: University of Toronto Press, 1971) Vol. 1, pp. 277-78.
2. M. Lescarbot, *Nova Francia - A description of Acadia 1606*, trans. P. Erondelle (London: George Routledge & Sons, 1928) p. 26.
3. N. Denys, *Description and natural history of Acadia*, ed. W. F. Ganong (Toronto: The Champlain Society, 1908) p. 203.
4. Ibid., p. 213.
5. A. H. Clark, *Acadia - The geography of early Nova Scotia to 1760* (Madison: University of Wisconsin Press, 1968) p. 141.
6. E. C. Wright, *The Petitcodiac* (Saint John: The Tribune Press, 1945) p. 10.
7. Clark, p. 145.
8. W. O. Raymond, *The River Saint John* (Saint John: The Tribune Press, 1943) pp. 67-68.
9. Wright, p. 14.
10. Ibid., p. 27.
11. E. C. Wright, *The Loyalists of New Brunswick* (Fredericton: 1955) p. 120.

References

12. E. Winslow, *The Winslow Papers*, ed. W. O. Raymond (Saint John: Sun Printing Co., 1901), p. 299.
13. W. S. MacNutt, *New Brunswick - A history 1784-1867* (Toronto: The Macmillan Co. of Canada Ltd., 1963) pp. 3-4.
14. W. O. Raymond, *The James White Papers* (Saint John: New Brunswick Historical Society, 1899-1905) Vol. 2, pp. 31-32.
15. *Royal Gazette and New Brunswick Advertiser*, Vol. 1 No. 68, January 23, 1787.
16. Wright, *The Loyalists of New Brunswick*, pp. 219-20.
17. Winslow, p. 354.
18. Ibid., pp. 493-94.
19. P. Fisher, *First history of New Brunswick* (Saint John: New Brunswick Historical Society, 1921 pp. 16-17. Originally published 1825.
20. Ibid., pp. 30-32.
21. *New Brunswick Royal Gazette*, Vol. 3, No. 4, April 1, 1817.
22. Ibid., Vol. 4, No. 45 January 5, 1819.
23. Ibid., Vol. 5, No. 51, February 15, 1320.
24. O. B. Bishop, *Publications of the Governments of Nova Scotia, Prince Edward Island and New Brunswick 1758-1952* (Ottawa: National Library of Canada, 1957) p. 154.
25. *Royal Gazette and New Brunswick Advertiser* Vol. 17, No. 926, May 19, 1802.
26. *New Brunswick Royal Gazette*, Vol. 6, No. 3, March 21, 1820.
27. Ibid., Vol. 6, No. 8, April 25, 1820.
28. Ibid., Vol. 6, No. 11, May 16, 1820.
29. Ibid., Vol. 6, No. 12, May 23, 1820.
30. Ibid., Vol. 12, No. 6, April 5, 1825.
31. Ibid., Vol. 12, No. 13, May 24,1825.
32. *New Brunswick Courier*, Vol. 15, No. 18, October 1, 1825.
33. Ibid.
34. Ibid.
35. *New Brunswick Royal Gazette*, Vol. 13, No. 11, May 9, 1826.
36. *New Brunswick Courier*, Vol. 15, No. 18, October 1, 1825.
37. *New Brunswick Royal Gazette*, Vol. 13, No. 51, February 27, 1827.
38. Ibid., Vol. 14, No. 30, September 25, 1827.

39. PANB, REX/pa Agricultural Societies 1846-47, Correspondence.
40. *Royal Gazette and New Brunswick Advertiser*, Vol. 14, No. 681, June 4, 1799.
41. *New Brunswick Royal Gazette*, Vol. 2, No. 8, April 30, 1816.
42. *Royal Gazette and New Brunswick Advertiser*, Vol. 1, No. 54, October 17, 1786.
43. Fisher, pp. 28-29.
44. *Royal Gazette and New Brunswick Advertiser*, Vol. 14, No. 646, Oct. 2, 1798.
45. P. Campbell, *Travels in the interior inhabited parts of North America in the years 1791 and 1792* (Toronto: The Champlain Society, 1937) p. 47.
46. J. F. W. Johnston, *Report on the agricultural capabilities of the Province of New Brunswick* (Fredericton: Queen's Printer, 1850) p. 161.
47. Ibid., p. 165.
48. PANB, REX/pa, Potatoe diseases 1845-46.
49. Fisher, p. 27.
50. The Acts of the General Assembly of Her Majesty's Province of New Brunswick 1786-1836 (Fredericton: Queen's Printer, 1836).
51. Johnston, p. 126.
52. Ibid., p. 70.
53. Ibid., p. 78.
54. Ibid., p. 171.
55. J. Robb, *An outline of improvement in agriculture, considered as a business, an art, and a science, with special reference to New Brunswick* (Fredericton: Queen's Printer, 1856) p. 16.
56. Ibid., p. 23.

Chapter IV — Prince Edward Island

1. D. C. Harvey, *The French regime in Prince Edward Island* (New Haven: Yale University Press, 1926) p. 16.
2. Ibid., p. 23.
3. A. H. Clark, *Three centuries and the Island* (Toronto: University of Toronto Press, 1959) p. 27.
4. Harvey, p. 37.

References

5. A. B. Warburton, *A history of Prince Edward Island* (Saint John: Barnes & Co. Ltd., 1923) p. 26.
6. Harvey, p. 45.
7. NA, MG1 G1 466/43.
8. NA, MG1 G1 466/47.
9. Harvey, pp. 63-64.
10. NA, MG1 C11 B, Vol. 11, p. 102.
11. Harvey, p. 105.
12. Ibid., p. 107.
13. Ibid., pp. 168-171.
14. Ibid., p. 180.
15. J. A. Caven, "A journey from Port La Joie to St. Peters 1751 (English summary of Franquet's report)," *P.E.I. Magazine*, 1900.
16. NA, MG1 C11 B, Vol. 33, p. 284.
17. Harvey, p. 86.
18. Ibid., p. 44.
19. NA, MG1 G1 466/43.
20. NA, MG1 G1 466/45-47.
21. Harvey, p. 208.
22. Sieur de la Roque, "Tour of inspection made by Sieur de la Roque," *Census 1752* (Ottawa: King's Printer, 1906) Vol. 2, Appendix A, Part 1 in report concerning Canadian Archives for the year 1905.
23. Ibid.
24. NA, MG1 C11 B, Vol. 33, p. 284.
25. NA, MG1 C11 B, Vol. 10, p. 90.
26. NA, MG1 C11 B, Vol. 20, p. 37.
27. Harvey, p. 90.
28. la Roque, p. 151.
29. NA, MG1 C11 B, Vol. 19, p. 12.
30. Harvey, p. 107.
31. NA, MG1 C11 B, Vol. 24, p. 193.
32. de la Roque, p. 151.
33. NA, MG1 C11 B, Vol. 28, p. 158.
34. de la Roque, p. 151.

35. NA, MG1 C11 B, Vol. 38, p. 269.

36. S. Holland, "Description of Prince Edward Island 1765," *Prince Edward Island Magazine* (Charlottetown: 1901) Vol. III, No. 4 p. 123.

37. NA, Colonial Office papers 226/1, pp. 11-12.

38. Clark, p. 67.

39. T. Curtis, "Voyage to the island of St. John 1775," *Journeys to the Island of St. John (P.E.I.)* ed. D. C. Harvey, (Toronto: The Macmillan Co. of Canada Ltd., 1955) p. 38.

40. Ibid., p. 39.

41. Ibid., p. 41.

42. W. Johnstone, "'Letters' and 'Travels' Prince Edward Island," *Journeys to the Island of St. John (P.E.I.)*, p. 126.

43. J. L. Lewellin, "Emigration," *Journeys to the Island of St. John (P.E.I.)*, p. 193.

44. Johnstone, pp. 96-97.

45. Ibid., pp. 129-130.

46. Ibid., p. 156.

47. Ibid., p. 96.

48. Lewellin, p. 206.

49. Johnstone, pp. 109-111.

50. *Prince Edward Island Register*, Vol. 1, No. 7, September 6, 1823.

51. Ibid., Vol. 1, No. 6, August 30, 1823.

52. Lewellin, pp. 210-212.

53. *Prince Edward Island Register*, Vol. 4, No. 163, February 27, 1827.

54. Ibid., Vol. 5, No. 193, September 25, 1827.

55. Ibid., Vol. 5, No. 198, November 6, 1827.

56. Ibid., Vol. 5, No. 204, December 18, 1827.

57. *Journal of the Legislative Assembly* (Charlottetown: King's Printer, 1834).

58. *Prince Edward Island Register*, Vol. 7, No. 317, February 23, 1830.

59. *Royal Gazette*, Vol. 7, No. 384, December 12, 1837.

60. Ibid., Vol. 10, No. 492, January 14, 1840.

61. Ibid., Vol. 7, No. 375, October 10, 1837.

62. Ibid., Vol. 11, No. 544, January 12, 1841.

63. Ibid.
64. Ibid., Vol. 8, No. 405, May 8, 1838.
65. Ibid., Vol. 11, No. 544, January 12, 1841.
66. Ibid., Vol. 10, No. 535, November 10, 1840.
67. Ibid., Vol. 7, No. 375, October 10, 1837.
68. Ibid., No. 380, November 14, 1837.
69. Ibid., Vol. 10, No. 496, February 11, 1840.
70. Ibid., Vol. 8, No. 388, January 9, 1838.
71. Ibid., No. 406, May 15, 1838.
72. NA, Colonial Office papers 228/8, p. 475.
73. NA, Colonial Office papers 226/1, pp. 11-12.
74. *Royal Gazette*, Vol. 15, No. 753, January 14, 1845.
75. Ibid., Vol. 8, No. 392, February 6, 1838.
76. Ibid., Vol. 15, No. 788, September 16, 1845.
77. Ibid., No. 789, September 23, 1845.
78. Johnstone, p. 134.
79. *Royal Gazette*, Vol. 15, No. 793, October 21, 1845.
80. *Royal Gazette*, Vol. 1, No. 4, August 24, 1854.
81. Johnstone, p. 128.
82. The acts will be found in the Journal of the Legislative Assembly for the pertinent years. NA, Colonial Office papers 228/1 through 228/9.

Chapter V — Quebec

1. M. Trudel, *The beginnings of New France 1524-1663*, trans. P. Claxton (Toronto: McClelland & Stewart Ltd., 1973) p. 40.
2. Ibid., p. 44.
3. S. Champlain, *The works of Samuel de Champlain*, ed. H. P. Biggar (Toronto: The Champlain Society 1922-26). Reprinted by University of Toronto Press, Toronto, 1971, Vol. 2, p. 44.
4. Ibid., Vol. 2, p. 52.
5. Ibid., Vol. 4, p. 337.
6. Ibid., Vol. 2, p. 147.
7. Trudel, p. 125.

8. Champlain, Vol. 2, p. 55.
9. F. Letourneau, *Histoire de l'agriculture (Canada francais)* (Montreal: L'Imprimerie, 1950) pp. 56 & 58.
10. M. Trudel, *Histoire de la Nouvelle-France. Le Comptoir 1604-1627* (Montreal: Fides, 1966) p. 322.
11. Champlain, Vol. 5, p. 112.
12. Father G. Sagard, *The long journey to the country of the Hurons*, trans. H. H. Langton, ed. G. M. Wrong (Toronto: The Champlain Society, 1939) p. 50.
13. Champlain, Vol. 3, p. 205.
14. Ibid., Vol. 6, p. 48.
15. Ibid., Vol. 5, pp. 256-257.
16. Ibid., Vol. 2, p. 44.
17. Trudel, *The beginnings of New France*, p. 127.
18. Champlain, Vol. 5, p. 203.
19. R. G. Thwaites, *The Jesuit relations and allied documents* (Cleveland: Burrows Bros. Co., 1896-1901) Vol. 6, pp. 49 & 73-75.
20. Ibid., Vol. 9, p. 163.
21. Ibid., Vol. 6, p. 29.
22. Ibid., Vol. 18, p. 85.
23. Ibid., Vol. 5, pp. 183, 215.
24. Ibid., Vol. 23, p. 271.
25. W. J. Eccles, *France in America* (New York: Harper & Row, 1972) p. 48.
26. Dollier de Casson, *History of Montreal 1640-1672*, trans. R. Flenley (J.M. Dent & Sons) p. 115.
27. Ibid., p. 123.
28. Ibid., p. 143.
29. Eccles, p. 57.
30. G. Lanctot, *A History of Canada: From its origin to the Royal Regime 1663*, trans. J. Hambleton (Toronto: Clarke, Irwin & Co., 1963) Vol. I, p. 154.
31. J. W. Eccles, *Canada under Louis XIV 1663-1701* (Toronto: McClelland & Stewart Ltd., 1964) p. 208.
32. L. A. Lahontan, *New Voyages to North America*, ed. R. G. Thwaites (Chicago: A.C. McClurg & Co., 1905) p. 37.

References

33. V. C. Fowke, *Canadian agricultural policy - The historical pattern* (Toronto: University of Toronto Press, 1945) p. 21.
34. NA, MG1 F3, Vol. 2, November 2, 1671.
35. NA, Report for 1899, Supplement, p. 115.
36. Ibid., p. 452.
37. Ibid., p. 469.
38. E. Salone, *La colonisation de la Nouvelle-France* (E. Guilmote Paris, 1906). Reprinted Boreal Express, Trois Rivieres, 1970, p. 378.
39. P. Kalm, *The America of 1750 - Peter Kalm's travels in North America* (New York: Dover Publcations Inc., 1966) Vol. 2, p. 438.
40. Eccles, *Canada under Louis XIV*, p. 16.
41. Ibid., p. 207.
42. Salone, pp. 374-375.
43. Kalm, Vol. 1, p. 383.
44. Ibid., p. 400.
45. Ibid., p. 401.
46. Ibid., Vol. 2, p. 438.
47. Ibid., p. 458.
48. Ibid., p. 476.
49. Ibid., p. 479.
50. Ibid., p. 493.
51. Ibid., p. 510.
52. Ibid., p. 514.
53. Ibid., pp. 529, 530, 531.
54. General J. Murray, *Report of the state of Government of Quebec in Canada June 5, 1762* (Quebec: Dessault & Proulx, 1902) pp. 51-52.
55. F. Ouellet, *Histoire Economique et Sociale du Quebec 1760-1850* (Montreal: Fides, 1966) pp. 46-52.
56. Murray, pp. 52-53.
57. Ibid., p35.
58. P. Campbell, *Travels in the interior inhabited parts of North America in the years 1791 and 1792* (Toronto: The Champlain Society, 1937) pp. 117-118.

59. I. Weld, *Travels through the states of North America and the provinces of Upper and Lower Canada during the years 1795, 1796, and 1797* (London: John Stockdale, 1807) Vol. 2, p. 7.

60. *Quebec Gazette*, April 23, 1789.

61. Ibid., December 31, 1789.

62. Ibid., March 11, 1790.

63. Ibid., April 15, 1750.

64. Weld, Vol. 1, p. 377.

65. *Quebec Gazette*, June 23, 1791.

66. Ibid., April 26, 1796.

67. R. L. Jones, "French-Canadian agriculture in the St. Lawrence Valley 1815-1850," *Agricultural History*, 16:137-148, July 1942, p. 147.

68. Campbell, pp. 113-114.

69. J. Lambert, *Travels through Canada and the United States of America in the years 1806, 1807, & 1808* (London: C. Cradock and W. Joy, 1814) Vol. 1, pp. 131-132.

70. Ouellet, p. 480.

71. J. Sansom, *Sketches of Lower Canada: Historical and descriptive 1817* (New York: Kirk & Mercein, 1817) p. 248.

72. J. Palmer, *Journal of travels in the United States of North America and Lower Canada - performed in the year 1817* (London: Sherwood, Neeley, and Jones, 1818) p. 209.

73. J. Bouchette, *General report of an official tour through the new settlements of the Province of Lower Canada - performed in the summer of 1824* (Quebec: Thomas Cary & Co., 1825) p. 81.

74. *Quebec Gazette*, December 1, 1791.

75. Ibid., August 31, 1805.

76. Ibid., July 14, 1806.

77. Ibid., August 4, 1806.

78. Ibid., September 14, 1807.

79. Ibid., January 4, 1808.

80. *Canadian Courant and Montreal Advertiser*, May 8, 1809.

81. The provincial statutes of Lower Canada (Quebec: King's Printer, 1811) Vol. 4-6, p. 122.

82. Jones, pp. 141-142.

83. Ibid., p. 143.

84. "Report of the Special Committee on the State of Agriculture in Lower Canada," *Journals of the Legislative Assembly of the Province of Canada* (Quebec: King's Printer, 1850) Vol 9, Appendix TT.

85. *Report of the Special Committee to whom were referred the report of the Lower Canada Agricultural Society and the special report of the Agricultural Society of the County of Beauharnois* (Quebec: King's Printer, 1852).

86. The ordinances from the French regime are found under the pertinent years in NA, MG1, Series F, Vol. 3. Ordinances in the English regime are from Appendix E, *Report of the Public Archives of Canada for 1913* (Ottawa: King's Printer, 1913). Proclamations are from Appendix B of the *Report of the Public Archives of Canada for 1921* (Ottawa: King's Printer, 1922). Acts are from *The Provincial Statutes of Lower Canada* (Quebec: King's Printer).

Chapter VI — Ontario

1. R. G. Thwaites, (ed.) *The Jesuit relations and allied documents* (Cleveland: Burrows Bros. Co., 1896-1901) Vol. 13, p. 93.

2. Ibid., Vol. 13, p. 101.

3. Ibid., Vol. 15, p. 137.

4. Ibid., p. 159.

5. Ibid., Vol. 28, p. 187.

6. Ibid., p. 229.

7. Ibid., Vol. 29, p. 221.

8. Ibid., Vol. 33, p. 255.

9. Ibid., p. 27.

10. W. J. Eccles, *The Canadian frontier 1534-1760* (New York: Holt, Rinehart & Winston, 1959) p. 46.

11. E. E. Rich, (ed.) *Minutes of the Hudson's Bay Company 1671-74* (Toronto: The Champlain Society, 1942) pp. 106-107.

12. Ibid., p. 108.

13. E. E. Rich, (ed.) *Copy book of letters outward 1680-1687* (Toronto: The Champlain Society, 1948) p. 9.

14. Ibid., p. 29.

15. E. E. Rich, (ed.) *Minutes of the Hudson's Bay Company 1679-84* (Toronto: The Champlain Society, 1946) Second Part 1682-84, pp. 295 & 308.

16. E. E. Rich, (ed.) *Minutes of the Hudson's Bay Company 1679-84* (Toronto: The Champlain Society, 1945) First Part 1679-82, Appendix A, p. 247.

17. Ibid., p. 251.
18. Ibid., p. 297.
19. E. E. Rich, (ed.) *Copy book of letters outward 1683-96* (London: The Hudson's Bay Record Society, 1957) p. 230.
20. Ibid., p. 273.
21. Hudson's Bay Company Correspondence inward, NA, Hudson's Bay Company papers MG20 A11/2 1703.
22. Ibid., 1706.
23. Albany Fort Journal, NA, Hudson's Bay Company papers, MG20 B3/a/1 1705-06.
24. Ibid.
25. E. E. Rich, (ed.) *Moose Fort Journals 1783-85* (London: The Hudson's Bay Record Society, 1954) p. 17.
26. Ibid.
27. Ibid., pp. 84-85.
28. J. Tod, *History of New Caledonia and the Northwest coast*, NA, Bancroft Collection MG29 C15, Vol. 3, C27, pp. 38-40.
29. Father L. Hennepin, *A new discovery of a vast country in America*, ed. R. G. Thwaites (Chicago: A.C. McClurg & Co., 1903) p. 45.
30. R. L. Jones, *History of agriculture in Ontario* (Toronto: University of Toronto Press, 1946) pp. 11-12.
31. G. P. de T. Glazebrook, *Life in Ontario* (Toronto: University of Toronto Press, 1968) pp. 17-18.
32. G. E. Reaman, *A history of agriculture in Ontario* (Don Mills: Saunders of Toronto Ltd., 1970), Vol. 1, p. 16.
33. L. J. Burpee, *Journals and letters of La Verendrye and his sons* (Toronto: The Champlain Society, 1927) p. 95.
34. Ibid., p. 101.
35. Ibid., p. 142.
36. Alexander Henry, *Travels and adventures in Canada and the Indian Territories between the years 1760 and 1776*, ed. J. Bain (Hurtig Ltd., 1959) p. 48.
37. Ibid., p. 56.
38. Ibid., p. 219.
39. Alexander Mackenzie, *The journals and letters of Sir Alexander Mackenzie*, ed. K. Lamb (Toronto: The Macmillan Co. of Canada Ltd., 1970) p. 98.

References

40. M. E. Campbell, *The Northwest Company* (Toronto: The Macmillan Co. of Canada Ltd., 1957) p. 138.
41. Ibid., p. 160.
42. "A survey of the settlement at Niagara, 25 August, 1782," *Public Archives Report of 1891*, Canadian Archives Series B, Vol. 169, p. 1.
43. E. A. Talbot, *Five years residence in the Canadas including a tour of the United States of America in the year 1823* (London: Longman, Hurst, Rees, Orme Brown and Green, 1824), Vol 2. Facsim reprint 1968 in Vol 1, p. 159.
44. P. Shirreff, *A tour of North America; together with a comprehensive view of the Canadas and the United States as adapted for agricultural emigration* (Edinburgh: Oliver & Boyd, 1835) p. 359.
45. S. Strickland, *Twenty-seven years in Canada West or the experience of an early settler* (London: Richard Bentley, 1853) p. 163.
46. Jones, p. 17.
47. E. C. Guillet, *The pioneer and backwoodsman* (Toronto: The Ontario Publishing Co. Ltd., 1963), Vol. 1, p. 312.
48. Strickland, p. 167.
49. Shirreff, p. 370.
50. Ibid., pp. 116-117.
51. Ibid., p. 120.
52. J. Howison, *Sketches of Upper Canada* (London: G. & W. B. Whittaker, 1821) p. 235.
53. W. Simpson, Letter from York, Upper Canada, July 14, 1832, NA Library.
54. Strickland, pp. 168-169.
55. Ibid., p. 158.
56. Shirreff, p. 213.
57. Ibid., p. 370.
58. P. Campbell, *Travels in the interior parts of North America in the years 1791 and 1792* (Toronto: The Champlain Society, 1937) p. 149.
59. Anonymous, "A letter from a gentleman to his friend in England, dated 20th November 1794" (Philadelphia: W. W. Woodward, 1795) p. 11.
60. Ibid., p. 15.
61. Ibid., p. 8.

62. I. Weld, *Travels through the states of North America and the provinces of Upper and Lower Canada 1795, 1796, 1797* (London: J. Stockdale, 1807) Vol. 2, p. 197.

63. Ibid., Vol. 2, p. 181.

64. G. Heriot, *Travels through the Canadas (1806)* (London: Richard Phillips, 1807) Vol. 1, p. 136.

65. Ibid., p. 151.

66. Jones, p. 75.

67. Strickland, p. 146.

68. C. P. Trail, *The backwoods of Canada* (London: Charles Knight, 1836). Republished by Coles Publishing Co., Toronto, 1971, p. 186.

69. W. Dunlop, *Tiger Dunlop's Upper Canada Comprising recollections of the American war of 1812-14 and statistical sketches of Upper Canada for the use of emigrants* (Toronto: McClelland and Stewart Ltd., 1967) p. 128.

70. Talbot, pp. 157-159.

71. Howison, pp. 135-136.

72. Talbot, pp. 299-302.

73. Ibid., pp. 177-180.

74. Ibid., p. 178.

75. *Upper Canada Gazette,* May 16, 1793.

76. J. J. Talman, *Agricultural Societies of Upper Canada* (Ontario Historical Society, 1931) Papers and Records Vol. 27, pp. 545-552.

77. *Upper Canada Gazette*, February 15, 1806.

78. Ibid., December 23, 1807.

79. Ibid., February 12, 1808.

80. F. V. Robinson, *A survey of agriculture* (thesis) (London: The University of Western Ontario, 1939) p. 322.

81. Guillet, Vol. 2 p. 129.

82. *Upper Canada Herald*, November 14, 1829.

83. *Colonial Advocate*, April 18, 1833.

84. Ibid., October 3, 1833.

85. Ibid., April 18, 1833.

86. Talman, p. 548.

87. Shirreff, p. 145.

88. *Kingston Gazette*, May 5, 1812.

References

89. *Upper Canada Gazette*, February 12, 1803.
90. *Colonial Advocate*, April 15, 1830.
91. *Upper Canada Gazette*, December 1, 1804.
92. *York Weekly Post*, December 21, 1821.
93. *Colonial Advocate*, July 2, 1829.
94. Ibid., March 1, 1832.
95. Shirreff, p. 117.
96. *Upper Canada Gazette*, May 11, 1805.
97. Ibid., January 2, 1817.
98. Talbot, p. 239.
99. Shirreff, p. 124.
100. Jones, p. 25.
101. H. Y. Hind, "Essay on the insects and diseases injurious to the wheat crop," *Journal of Agriculture of Upper Canada*, 1858, Appendix, pp. 23. 39-41.
102. *Upper Canada Gazette*, November 20, 1802.
103. Heriot, p. 133.
104. Reaman, p. 28.
105. Jones, p. 101.
106. *British American Cultivator*, Vol. 3, No. 10, October 1844, p. 150.
107. Strickland, p. 302.
108. Talbot, p. 166.
109. *Canadian Agriculturist*, Vol. 1, No. 6, June 1, 1849.
110. Ibid., Vol. 1, No. 11, November 1849.
111. J. Logan, *Notes of a journey through Canada, the United States of America and the West Indies* (Edinburgh: Fraser & Co., 1838) pp. 48-49.
112. Jones, p. 96.
113. *Upper Canada Gazette*, April 9, 1818.
114. Journals of the Legislative Assembly and House of Assembly of Upper Canada for the pertinent years.

Chapter VII — Manitoba

1. E. E. Rich, (ed.) *Hudson's Bay Company copy book of letters, commissions, instructions outward 1688-1696* (London: The Hudson's Bay Record Society, 1957) pp. 195-196.

2. *York Factory Journal* (1715-16), NA, Hudson's Bay Company papers MG20 B239/a/2, September 14, 17, 19, & 20, 1715.

3. Ibid., (1716-17), B239/a/3, September 22 & October 4, 1716.

4. *Churchill Post Journal* (1732-33) NA, Hudson's Bay Company papers MG20 B42/a/13, April 9 & 23, 1733.

5. Ibid., (1733-34), B42/a/14, May 13, 1734.

6. Rich, pp. 194-195.

7. Ibid., p. 245.

8. *York Factory Journal* (1714-15) B239/a/1, October 18, 1714 & May 9, 1715.

9. Ibid., (1715-16) B239/a/2, September 29 & October 5, 1715.

10. Ibid.

11. K. G. Davies, *Letters from Hudson Bay 1703-1740* (London: The Hudson's Bay Record Society, 1965) p. 162.

12. Ibid., p. 175.

13. *Churchill Post Journal* (1732-33), B42/a/13, August 5, 1732, March 2 & May 26, 1733.

14. Ibid., (1733-34), B42/a/14, August 18 & 20, September 22, 1733.

15. Ibid., (1732-33) B42/a/13, August 5, 1732, March 2 & May 26, 1733.

16. Ibid., (1733-34) B42/a/14, August 18 & 20, September 22, 1733.

17. Ibid., (1732-33) B42/a/13 August 5, 1732, March 2 & May 26, 1733.

18. *Churchill Post* Invoices, NA, Hudson's Bay Company papers, MG20 A24/17 for the years 1733, 34, 35, 36, 38, & 39.

19. *York Factory Journal* (1752-53), B239/a/36, October 13 & 20, 1752.

20. Ibid.

21. Ibid.

22. Ibid., (1774-75), B239/a/71, September 29, 1774; B239/a/7s, October 9, 11, 15, 20, 28, & December 18, 1777.

23. Ibid.

24. Ibid.

25. Ibid.

26. Ibid.

References

27. Ibid., (1759-60), B239/a/47, May 27, 1760.

28. Ibid., (1774-75), B239/a/71 September 29, 1774; B239/a/7s, October 9, 11, 15, 20, 28, & December 18, 1777.

29. *Brandon House Journal*, NA, Hudson's Bay Company papers, MG20 B22/a/2, May 4, 1795; a/4, September 27, 1796; a/5, October 8, 1797; a/6, October 3, 1798; a/8, October 7, 1800; a/14, 0ctober 16, 1806; a/15, October 7, 1807 & April 21, May 5 & 7, 1808; a/17, September 14 & October 10, 1809.

30. Ibid.

31. Ibid.

32. Ibid.

33. Ibid.

34. Ibid.

35. Ibid.

36. Ibid.

37. Ibid.

38. Ibid.

39. Ibid.

40. E. Coues, (ed.) "New light on the early history of the greater Northwest," *The manuscript journal of Alexander Henry (the younger) and of David Thompson* (New York: F. H. Harper, 1897). Repritned by Ross & Haines, Minneapolis, 1965, Vol. 2, p. 228.

41. Ibid., Vol. 1, p. 291.

42. D. W. Harmon, *Sixteen years in the Indian country*, ed. W. K. Lamb (Toronto: The Macmillan Co. of Canada Ltd., 1957) p. 34.

43. NA, Selkirk Papers MG19 E1, Vol. 3, p. 782.

44. *Oxford Post Journal* (1810-11), NA, Hudson's Bay Company papers, MG20 B156/a/4, July 26, 1811.

45. G. Franchere, "Narrative of a voyage to the Northwest coast of America in 1811, 1812, 1813, and 1814," *Early western travels*, ed. R. G. Thwaites (Cleveland: The A.H. Clarke Co., 1904-07) Vol VI, p. 331.

46. A. Amos, *Report of the trials in the courts of Canada, relative to the destruction of the Earl of Selkirk's Settlement on the Red River* (London: John Murray, 1820) p. 6.

47. NA, Selkirk Papers, p. 766.

48. Ibid., pp. 788-89.

49. Ibid., p. 955.

50. A. Ross, *The Red River settlement, its rise, progress, and present state* (London: Smith, Elder and Co., 1856) pp. 50-51.

51. F. H. Scholfield, *The story of Manitoba* (Winnipeg: The S.J. Clarke Co.) Vol. 1, pp. 156-157.

52. NA, Selkirk Papers, Vol. 4, p. 1066.

53. J. W. G. MacEwan, *Blazing the old cattle trails* (Saskatoon: Modern Press, 1962) pp. 14-19.

54. NA, Selkirk Papers, Vol. 2, p. 394.

55. Ibid., Vol. 3, p. 768.

56. J. W. Todd, Jr. (ed.) *Two journals of Robert Campbell 1808 to 1853.* (Seattle: 1958) pp. 11-26.

57. W. L. Morton, *Manitoba - a history* (Toronto: University of Toronto Press, 1957) p. 61.

58. A. H. R. Buller, *Essays on wheat* (New York: The Macmillan Co., 1919) p. 16.

59. E. H. Oliver, *The Canadian North-West: Its early development and legislative records* (Ottawa: Government Printing Bureau, 1914) Public Archives of Canada Pub., No. 9, Vol. 1, p. 74.

60. Ibid., p. 663.

61. Ibid., Vol. 2, p. 870.

62. W. A. Mackintosh, *Prairie settlement - The geographical setting* (Toronto: The Macmillan Co. of Canada Ltd., 1934) p. 27.

63. Ibid., p. 46.

64. Ross, p. 213.

65. Oliver, Vol. 1, p. 264.

66. Ibid., p. 265.

67. Ibid., p. 292.

68. Ibid., p. 320.

69. Ibid., pp. 85-86.

70. Ibid., p. 358.

71. Ross, p. 389.

72. Ibid., p. 390.

73. Todd, pp. 26-27.

74. Ross, p. 392.

75. Ibid., p. 390.

76. Morton, p. 116.

References

77. *The Manitoban*, April 29, 1871.
78. J. W. G. MacEwan, *Between the Red and the Rockies* (Toronto: University of Toronto Press, 1952) p. 51.
79. *The Manitoban*, July 6, 1872.
80. Ibid., May 2, 1874.
81. A. Begg, *History of the North-West* (Toronto: Hunter, Rose & Company, 1894) pp. 387-388.
82. *Saskatchewan Herald*, August 25, 1878.
83. Morton, pp. 158-159.
84. Ibid., p. 177.
85. *Daily Free Press*, January 22, 1875.
86. J. C. Southesk, *Saskatchewan and the Rocky Mountains* (Toronto: James Campbell & Son, 1875 and Edinburgh: Edmonston and Douglas, 1875) p. 358.
87. Morton, p. 145.
88. Ibid., p. 180.
89. S.N. Murray, *The valley comes of age* (Fargo: North Dakota Institute of Regional Studies, 1967) p. 39.
90. *The Nor'Wester*, April 28, 1860.
91. *The Manitoban*, October 15, 1870.
92. Ibid., July 1, 1871.
93. *Manitoba Free Press*, July 11, 1874.
94. "Evidence before the Select Committee on Immigration and Colonization 1876," Province of Manitoba and North West Territory of the Dominion of Canada (Ottawa: Department of Agriculture, 1876).
95. *The Manitoban*, May 13, 1871.
96. Ibid., June 3, 1871.
97. *Manitoba Free Press*, December 27, 1873.
98. Ibid., October 23, 1876.
99. Ibid., October 17, 1877.
100. J. G. McGregor, *Edmonton Trader - The story of John A. McDougall* (Toronto: McClelland & Stewart Ltd., 1963) p. 85.
101. H. G. L. Strange, *A short history of prairie agriculture* (Winnipeg: Searle Grain Co., 1954) p. 39.
102. *The Nor'Wester*, October 9 1862.

103. *Manitoba Free Press*, March 21, 1874.
104. Ibid., June 6, 1874.
105. The information in this paragraph was obtained from numerous issues of *The Manitoban*, 1871-1873.
106. *Manitoba Free Press*, July 26, 1873.
107. *Daily Free Press*, January 20, 22 & February 4, 1875.
108. Ibid., July 29, 1874.
109. *The Nor'Wester*, February 6, 1865.
110. Ibid.
111. Ibid., March 31, 1864.
112. Ibid., April 14, 1860.
113. *The Manitoban*, April 20, 1872.
114. *The Nor'Wester*, August 28, 1860.
115. *Nor'West Farmer*, August 1889.
116. Oliver, Vol. 2, p. 1120.
117. Statutes of Manitoba.

Chapter VIII — Saskatchewan

1. K. Rasmussen, "The myth that La Corne grew wheat in 1754," *Saskatchewan History*, Vol. XXXIII, No. 1, Winter 1980.
2. *Cumberland House Post Journal*, NA, Hudson's Bay Company papers MG20 B49/a/4, October 25, 1776.
3. Ibid., B49/a/14, May 7 & 11, 1784.
4. *Hudson House Post Journal*, NA, Hudson's Bay Company papers MG20 B87/a/6, May 10 & 12, 1783.
5. *Cumberland House Journal*, B49/a/24, May 9, 1793.
6. *South Branch House Journal*, NA, Hudson's Bay Company papers MG20 B205/a/3, October 1, 1788 and May 8, 1789.
7. Ibid., B121/a/4, May 27 & October 12, 1789.
8. Ibid.
9. *Manchester House Post Journal*, NA, Hudson's Bay Company papers MG20 B121/a/8, October 23, 1792.

References

10. *Cumberland House Journal*, B49/a/28, Dates as in the Ms.
11. Ibid.
12. Ibid., B49/a/32a, September 25, 1802.
13. Ibid., B49/a/34, September 23 to October 7, 1818.
14. *Carlton House Post Journal*, NA, Hudson's Bay Company papers MG20 B27/a/4, April 26, 1815.
15. Ibid., B27/a/4, November 1, 1814.
16. Ibid., B27/a/8, May 13, 1819.
17. *Cumberland House Journal*, B49/a/34, April 23, 1819.
18. *Carlton House Journal*, B27/a/4, May 11, 1815.
19. Ibid., B27/a/4, May 3, 1815.
20. *Cumberland House Journal*, B49/a/34, May 8, 1819.
21. Ibid., B49/a/35, September 6, 1819.
22. J. Franklin, *Narrative of a journey to the shores of the Polar Sea in the years 1819-20-21-22* (London: John Murray, 1824) p. 180.
23. *Cumberland House Journal*, B49/a/35, October 12, 1819.
24. *Carlton House Journal*, B27/a/4, April 29, 1815.
25. D. W. Harmon, *Sixteen years in the Indian country*, ed. W. K. Lamb (Toronto: The Macmillan Co. of Canada Ltd., 1957) p. 83.
26. *Carlton House Journal*, B27/a/5, June 22, 1815.
27. Harmon, p. 61.
28. *Carlton House Journal*, B27/a/5, July 20, 1815.
29. "Letter from J. Leith to John Macleod, August 27, 1825," NA, John Macleod papers (1811-37) MG19 A23.
30. Harmon, p. 85.
31. *Carlton House Journal*, B27/a/13, September 5 & 11, 1823.
32. Franklin, p. 86.
33. *Cumberland House Journal*, B49/a/35, August 25, 1819.
34. Ibid., B49/a/35, May 16, 1820.
35. Ibid., B49/a/35, October 29, 1819.
36. *Carlton House Journal*, B27/a/8, October 10, 1818.
37. J. McLean, *Notes of a twenty five year's service in the Hudson's Bay Territory*, ed. W. S. Wallace (Toronto: The Champlain Society, 1932) p. 134.
38. Ibid.

39. *The Manitoban*, March 4, 1871.

40. G. M. Grant, *Ocean to ocean* (Toronto: The Radisson Society of Canada, 1925) p. 15341.

The Manitoban, June 7, 1873.

42. T. Spence, *The Saskatchewan country of the North-West of the Dominion of Canada* (Montreal: Lovell Printing and Publishing Company, 1877) p. 9.

43. Ibid., p. 11.

44. A. McPherson, *The Battlefords - A history*, (Saskatoon: Modern Press, 1957) p. 39.

45. Ibid., p. 53.

46. *Saskatchewan Herald*, August 25, 1878.

47. Ibid., August 25, 1879.

48. *Prince Albert Times and Saskatchewan Review*, March 14, 1884.

49. *Regina Leader*, November 22, 1883.

50. W. H. Barneby, *Life and labour in the far, far west* (London: Cassell & Co. Ltd., 1884) p. 226.

51. *Regina Leader*, June 19, 1884.

52. *Prince Albert Times*, February 1, 1884.

53. This item appeared as an advertisement by the Department of the Interior in J. W. Powers, *History of Regina* (Regina: The Leader Co. Ltd., 1887) p. 25.

54. Ibid.

55. *Regina Leader*, January 3, 1884.

57. N. F. Black, *History of the Province of Saskatchewan and the Old North West* (Regina: North West Historical Co., 1913) p. 500.

57. *Nor West Farmer & Miller*, June 1889.

58. Black, p. 494.

59. *Prince Albert Times*, January 31, 1883.

60. Ibid., Aug. 27, 1886.

61. *Regina Leader*, January 17, 1884.

62. H. G. L. Strange, *A short history of prairie agriculture* (Winnipeg: Searle Grain Co., 1954) p. 39.

63. Spence, p. 9.

643. *Saskatchewan Herald*, September 9, 1878.

65. Ibid., September 27, 1880.

References

66. Spence, p. 11.
67. W. M. Elkington, *Five years in Canada* (London: Whittaker & Co., 1895) p. 46.
68. Ibid., p. 52.
69. *Regina Leader*, June 21, 1883.
70. J. Hawkes, *Saskatchewan and its People* (Chicago: The S. J. Clarke Publishing Co., 1924) p. 603.
71. Elkington, p. 63.
72. Grant, p. 153.
73. J. F. C. Wright, *Saskatchewan, the history of the province* (Toronto: McClelland & Stewart Ltd., 1955) p. 77.
74. Ibid., p. 99.
75. Hawkes, p. 600.
76. Elkington, p. 64.
77. *Regina Leader*, September 27, 1883.
78. Consolidated Ordinances of the North West Territories to 1898.
79. Ibid.
80. *Cumberland House Journal*, B49/a/35, October 12, 1819.
81. *Saskatchewan Herald*, July 19, 1880.
82. *Prince Albert Times*, April 2, 1889.
83. *Regina Leader*, January 17, 1884.
84. Ibid., January 3, 1884.
85. Ibid., January 27, 1885.
86. *Nor West Farmer & Miller*, May 1889.
87. Ibid., August 1889.
88. Ibid., May 1889.
89. Elkington, p. 52.
90. *Saskatchewan Herald*, June 16, 1879.
91. Ibid., June 7, 1880.
92. *Nor West Farmer & Miller*, June 1890.
93. Ibid., January 27, 1879.
94. *Regina Leader*, April 3, 1884.
95. Ibid., October 9, 1884.

96. Consolidated Ordinances of the North West Territories to 1898.
97. Wright, p. 111.
98. *Regina Leader*, February 14, 1884.
99. Consolidated Ordinances of the North West Territories to 1898.
100. *Regina Leader*, April 3, 1888.
101. Ibid., May 8, 1888.
102. Consolidated Ordinances of the North West Territories to 1898.

Chapter IX — Alberta

1. E. B. Swindlehurst, *Alberta agriculture - historical review* (Edmonton: Alberta Department of Agriculture, 1967) pp. 10-11.
2. A. Mackenzie, *The journals and letters of Sir Alexander Mackenzie*, ed. W. K. Lamb (Toronto: The Macmillan Co. of Canada Ltd., 1970) p. 242.
3. Ibid.
4. *Edmonton House Post Journal*, NA, Hudson's Bay Company papers MG20 B60/a/2, October 12, 1796.
5. Ibid., B60/a/3, May 4, 1798.
6. Ibid., May 7, 1798.
7. Ibid., B60/a/4, September 21, 1798.
8. Ibid., October 3, 1798.
9. Ibid., November 25, 1798.
10. L. R. Masson, *Le Bourgeois de la Compagnie du Nord-Ouest* (Quebec: De L'Imprimerie Generale A. Cote Cte., 1889-1890) Vol. 2, p. 386.
11. J. N. Wallace, *The wintering partners on Peace River from the earliest records to the union in 1821* (Ottawa: Thorburn & Abbott, 1929) p. 76.
12. Ibid., p. 122.
13. D. W. Harmon, *Sixteen years in the Indian country*, ed. W. K. Lamb (Toronto: The Macmillan Co. of Canada Ltd., 1957) p. 118.
14. Ibid., p. 121.
15. Ibid.
16. Ibid., p. 124.
17. Ibid., p. 127.
18. Ibid., p. 126.

References

19. E. Coues, (ed.) *New light on the early history of the greater northwest*, The manuscript journals of Alexander Henry (the younger) and of David Thompson (New York: Francis P. Harper, 1897) Vol. 2, p. 549.
20. Ibid., pp. 604-605.
21. Ibid., p. 609.
22. Ibid., p. 617.
23. Ibid., p. 620.
24. Ibid., p. 621.
25. Ibid., p. 628.
26. *Edmonton House Journal*, B60/a/10, October 18, 1811.
27. *Fort Vermilion Post Journal*, NA, Hudson's Bay Company papers MG20 B224/a/2, October 4 & 5, 1826.
28. Ibid.
29. Ibid., B224/a/2, September 23, 1826.
30. Ibid., B224/a/3, September 22, 1827.
31. *Edmonton House Journal*, B60/a/13, April 24 to May 3, 1815.
32. Ibid., B60 for the pertinent years.
33. "Letter from J. M. Stuart, January 5, 1827," NA, John McLeod papers (1811-1837) MG19 A23.
34. *Dunvegan Post Journal*, NA, Hudson's Bay Company papers MG20 B56/a/4, June 1, 1835.
35. Coues, p. 604-605.
36. *Fort Vermilion Journal*, B224/a/2, September 6, 1826.
37. *Edmonton House Journal*, B60/a/15, November 22, 1815.
38. Cous, p. 621.
39. *Edmonton House Journal*, B60/a/15, November 22, 1815.
40. Ibid., 860/a/26, June 21, 1828.
41. Ibid., B60/a/27, June 12, 1832.
42. Ibid., B60/a/28, June 15, 1833.
43. *Dunvegan Post Journal*, B56/a/4, June 12, 1845.
44. *Fort Vermilion Journal*, B224/a/7, August 15, 1840.
45. *Edmonton House Journal*, B60/a/22, September 25, 1823.
46. Ibid., B60/a/18, April 22, 1820.
47. Ibid., B60/a/27, June 5, 1832.

48. Ibid., July 9, 1832.
49. Ibid., October 22, 1832.
50. Ibid., November 30, 1832.
51. Ibid., December 3, 1832.
52. Ibid., B60/a/28, November 17, 1833.
53. Ibid., November 18, 1833.
54. *Fort Vermilion Journal*, B224/a/12, May 11, 1864.
55. Ibid., May 14, 1864.
56. *Edmonton House Journal*, B60/a/28, May 28, 1833.
57. *Dunvegan Post Journal*, B56/a/4, August 23, 1835.
58. Ibid., August 24, 1835.
59. Ibid., B56/a/9, October 7, 1839.
60. Ibid., May 5, 1840.
61. *Edmonton House Journal*, B60/a/24, October 26, 1826.
62. Ibid., December 1, 1826.
63. Ibid., B60/a/23, April 9, 1825.
64. Ibid., B60/a/24, September 11, 1826.
65. C. M. MacInnes, *In the shadow of the Rockies* (London: Rivington's, 1930) p. 237.
66. *Edmonton House Journal*, B60/a/27, April 29 & 30, 1833.
67. J. D. MacGregor, *Edmonton - a history* (Edmonton: M.G. Hurtig Publishers, 1967) p. 45.
68. *The Nor Wester*, March 9, 1867.
69. J. McDougall, *George Millard McDougall, pioneer patriot and missionary* (Toronto: William Briggs, 1902) p. 120.
70. Ibid., p. 105.
71. Ibid., p. 128.
72. MacGregor, *Edmonton - a history*, p. 65.
73. *The Saskatchewan Herald*, August 25, September 9, October 7, 1878.
74. Ibid.
75. Ibid., October 6, 1879.
76. *Edmonton Bulletin*, December 20, 1880.
77. Ibid., March 21, 1881.

References

78. Ibid., January 3, 1881.
79. Canada census, 1881.
80. *Edmonton Bulletin*, December 6, 1880.
81. Ibid., February 21, 1881.
82. Ibid., January 12, 1884.
83. Ibid., January 19, 1884.
84. Ibid., January 3 & February 21, 1881.
85. Ibid., December 6, 1880.
86. MacInnes, p. 99.
87. McDougall, p. 199.
88. L. V. Kelly, *The range men* (Toronto: William Briggs, 1913) p. 111.
89. Ibid., p. 113.
90. Letter from Bunns to Hardisty, August 14, 1875, from Bow River (Calgary: Glenbow Foundation).
91. E. Lynch-Staunton, *Ranching in Southern Alberta* (Calgary: Glenbow Foundation).
92. Fragments, (Calgary: Glenbow Foundation).
93. Ibid.
94. MacInnes, p. 194.
95. Kelly, p. 112.
96. *The Fort Macleod Gazette*, August 24 & October 4, 1882.
97. Ibid.
98. Ibid., July 1, 1882.
99. Ibid., October 4, 1882.
100. Kelly, pp. 115-116.
101. *Saskatchewan Herald*, May 23, 1881.
102. Ibid., October 3, 1881.
103. *Edmonton Bulletin*, November 19, 1881.
104. *Fort Macleod Gazette*, August 24, 1882.
105. Ibid., September 4, 1882.
106. *Lethbridge News*, February 5, 1886.
107. Ibid., May 21, 1886.
108. *Edmonton Bulletin*, March 4, 1882.

109. Ibid., April 28, 1883.

110. J. Blue, *Alberta, past and present* (Chicago: Pioneer Historical Publishing Co., 1924) Vol. 1, pp. 321-322.

111. Ibid.

112. *Calgary Weekly Herald*, April 11, 1888.

113. Blue, p. 328.

114. *Lethbridge News*, February 12, 1886.

115. *The Calgary Herald and Alberta Live Stock Journal*, August 29, 1888.

116. *The Fort Macleod Gazette*, September 14, 1882.

117. Ibid., June 4, 1883.

118. Ibid., October 24, 1882.

119. Ibid., July 14, 1883.

120. N. W. Macleod, *Picturesque Cardston and environment*, Cardston, N.W.T., 1900, pp. 7-8.

121. J. G. MacGregor, *A history of Alberta*, (Edmonton: M.G. Hurtig, 1972) p. 165.

122. *Calgary Herald*, December 4, 1889.

123. MacGregor, *A history of Alberta*, p. 167.

124. Blue, p. 213.

125. MacGregor, *A history of Alberta*, p. 165.

126. Blue, p. 353.

127. MacGregor, *A history of Alberta*, p. 165.

128. *The Fort Macleod Gazette*, May 14, 1883.

129. Blue, p. 357.

130. *Edmonton Bulletin*, April 4, 1881.

131. Ibid., April 29, 1882.

132. Ibid., February 10, 1883.

133. *Lethbridge News*, February 9, 1887.

134. Ibid., April 6, 1887.

135. Ibid., February 19, 1886.

136. *Edmonton Bulletin*, March 14, 1881.

137. *The Fort Macleod Gazette*, March 28, 1889.

138. Blue, p. 346.

139. *Edmonton Bulletin*, March 21, 1881.

140. *The Fort Macleod Gazette*, January 17, 1889.

141. Ibid., June 13, 1889.

142. Ibid., August 1, 1889.

143. *Lethbridge News*, September 20, 1886.

144. *Saskatchewan Herald*, June 30, 1879.

145. Ibid., September 27, 1880.

146. *Lethbridge News*, July 28, 1886.

147. *Edmonton Bulletin*, July 15, 1882.

148. Ibid., November 5, 1881.

149. Ibid., September 30, 1882.

150. *Lethbridge News*, May 14, 1886.

151. *The Fort Macleod Gazette*, October 4, 1888.

152. Ibid., August 16, 1887.

153. *Calgary Tribune*, February 8, 1888.

154. *Lethbridge News*, February 23, 1888.

155. Ibid., May 21, 1886.

156. Ibid., May 11, 1887.

Chapter X — British Columbia

1. G. P. V. Akrigg & B. Helen, *British Columbia chronicle 1778-1846* (Victoria: Discovery Press, 1975) p. 58.

2. G. Vancouver, *A voyage of discovery to the North Pacific Ocean and around the world*, ed. J. Vancouver (London: G.G. and J. Robinson, and J. Edwards, 1798) Vol. 1, p. 393.

3. D. W. Harmon, *Sixteen years in the Indian country*, ed. W. K. Lamb (Toronto: The Macmillan Co. of Canada Ltd., 1957) p. 139.

4. Ibid., p. 192.

5. Ibid., pp. 192-193.

6. Ibid., p. 176.

7. *Fort St. James Post Journal*, NA, Hudson's Bay Company papers MG20 B188/a/1, 1820.

8. Ibid., B188/a/2, 1823.

9. Ibid., B188/a/4, 1824.

10. *Fort Langley Post Journal*, NA, Hudson's Bay Company papers MG20 B113/a/1, 1827.

11. Ibid., B113/a/2, 1828.

12. Ibid., B113/a/3, 1829.

13. Journal of the Hudson Bay Company kept at Fort Langley during the years 1827-30, BCARS, Bancroft Collection MG29 C15, Vol. 3, C22, p. 116.

14. J. Work, *The Journal of John Work January to October 1835*, ed. H. D. Dee (Victoria: C.F. Banfield, 1945) p. 84.

15. J. McLean, *Notes of a twenty-five year's service in the Hudson's Bay Territory*, ed. W. S. Wallace (Toronto: The Champlain Society, 1932) p. 151.

16. Ibid., p. 161.

17. Ibid., p. 170.

18. R. Finlayson, *History of Vancouver and the Northwest Coast*, BCARS, MG29 C15, Vol. 2, p. 26.

19. Ibid., p. 28.

20. E. E. Rich, *The letters of John McLoughlin from Fort Vancouver to the Governor and Committee* (Toronto: The Champlain Society, 1941) Vol. 2, p. 216.

21. Work, p. 45.

22. *Fort Simpson Post Journal*, NA, Hudson's Bay Company papers MG20 B20/a/3, May 5 to October 12, 1835.

23. E. O. S. Scholefield, *British Columbia from the earliest times to the present* (Vancouver: S.J. Clarke Publishing Co., 1914) Vol. 1 p. 527.

24. Ibid., p. 511.

25. Ibid., p. 553.

26. A. Begg, *The history of British Columbia*, (Toronto: William Briggs, 1894) p. 284.

27. *The British Colonist*, December 11, 1858.

28. Ibid., February 12, 1859.

29. Ibid., March 5, 1859.

30. Ibid., June 13, 1859.

31. Ibid., March 19, 1859.

32. Ibid., July 15, 1859.

33. Scholefield, Vol. 2, p. 591.

34. *The British Columbian*, February 13, 1861.

References

35. Ibid., April 4, 1861.
36. Ibid., July 11, 1861.
37. Ibid., July 9, 1862.
38. Ibid., Jan.21, 1863.
39. R. C. Brown, *British Columbia - An essay* (New Westminster: The Royal Engineer Press, 1863) p. 6.
40. Ibid., p. 8.
41. Ibid., p. 13.
42. Ibid., p. 14.
43. Ibid., pp. 20-21.
44. Ibid., p. 36.
45. Ibid., p. 37.
46. Ibid., p. 38.
47. Ibid., pp. 40-41.
48. M. Macfie, *Vancouver Island and British Columbia: Their history, resources, and prospects* (London: Longman, Green, Longman, Roberts & Green, 1865) p. 196.
49. Brown, p. 41.
50. *The British Columbian* March 18, 1863.
51. Ibid., April 25, 1863.
52. Ibid., April 1, 1863.
53. Ibid., May 23, 1863.
54. Ibid., July 9, 1862.
55. Ibid., May 23, 1863.
56. Ibid., June 3, 1863.
57. Scholefield, Vol. 2, p. 593.
58. Macfie, pp. 201-203.
59. Rich, Vol. 1, p. 143.
60. Ibid., Vol. 2, p. 302.
61. *The British Columbian*, June 14, 1862 (citing *The British Colonist*).
62. Ibid., May 23, 1863 (citing *The British Colonist*).
63. Ibid., August 13, 1863.
64. *The British Colonist* May 5, 1869.

65. Ibid., June 7, 1869.
66. Ibid., May 8, 1860.
67. Ibid., September 2, 1869.
68. *The British Columbian*, November 13, 1861.
69. *The British Colonist*, August 4, 1869.
70. J. W. G. MacEwan, *Agriculture on parade, the story of the fairs and exhibitions of Western Canada* (Toronto: Thomas Nelson & Sons (Canada), 1950) p. 16.
71. *The British Colonist*, November 12, 1870.
72. Ibid., September 15, 1870.
73. *Fort St. James Post Journal*, 1820.
74. A. MacDonald, "Letter to John McLeod from Ft. Langley, January 15, 1832," NA, MG19 A23, p. 290.
75. Work, p. 84.
76. McLean, p. 151.
77. *Fort Langley Post Journal*, B113/a/3, July 13, 1829.
78. Ibid., B113/a/3, October 5, 8, 12, November 25, & December 19, 1829.
79. Ibid., B113/a/2, August 3, 1828.
80. *Fort Alexandria Journal*, B5/a/2, June 28 & August 3, 1827.
81. *The British Columbian*, July 2 & September 12, 1862.
82. C. W. Vrooman, "A history of ranching in British Columbia," *Economic Annalist*, Vol. XI, No. 2, April 1941, pp. 20-23.
83. Ibid., July 19, 1862.
84. Ibid., January 23, 1862.
85. *Fort St. James Post Journal*, 1823.
86. *The British Colonist*, March 26, 1870.
87. Ordinances and statutes of British Columbia for the pertinent years.
88. Ibid., July 10, 1870.
89. Macfie, p. 221.
90. *The British Columbian*, November 13, 1861.
91. Ibid., February 18, 1863.
92. *The British Colonist*, January 7, 1870.
93. Ibid., August 5, 1870.

References

94. Ibid., September 18 & 5, 1869.
95. Brown, p. 41.

The Trail Blazers of Canadian Agriculture

Index

A
aboteaux 41, 43
Agricola 41, 82, 115
agricultural
  extension and education 44, 46, 67, 118, 144
  incentives 123, 169
  policy 14-15, 100
  potential 16
  practices 142
  societies 15, 19, 22, 42, 44, 62, 63, 66, 67, 82, 86, 87, 90, 111, 113, 142, 143, 152, 179, 180, 204, 221, 230, 247
  subsidies 34, 35
animal control 47
apples 28, 37, 50, 96, 138, 243
auction sales 14

B
barley 2, 16, 35, 58, 60, 79, 99, 100, 107, 128, 131, 165, 187, 213, 215, 219, 234, 236, 237, 244
beans 2, 4, 30, 58
bears 79
Bell Farm 197-198
bird pests 182
black birds 176, 181, 192, 202, 203
blight 50, 91
buckwheat 60, 75, 165
buffalo meat 190

C
cabbages 16, 38, 75, 79, 162, 187, 237
cactus 243
Canadian Agricultural Coal and Colonization Co. 198
Canadian Pacific Railway 196, 253
cart trains 168
cattle
  general 8, 9, 10, 17, 18, 20, 31, 33, 38, 4l, 55, 58, 76, 99, 107, 129, 130, 131, 157, 158, 159, 160, 163, 166, 191, 217, 248
  diseases 229
  exports 225
  imports 176, 223, 224

The Trail Blazers of Canadian Agriculture

upgrading 224
census 20, 31, 32, 55, 56, 74, 75, 76, 77, 167
Central Agricultural Society 62-63
cereals 39
climate 18, 164, 181, 192, 226
Cochrane Ranch 223, 224
colonizing companies 197, 199
contract production 115
corporate farms 197
crops
    general 57, 58, 59, 75, 79
    failures 50, 117, 215
    rotations 245
    variety testing 89, 230
crown land 13
    sales 251
crows 181
cultivation tools 36
cultural
    methods 140, 157, 176
    problems 139, 235

D

dairy cattle 83, 222
dairy production 170, 177, 204, 239
demonstration farms 102
departments of agriculture 152
dogs 23, 160, 217, 249
drainage 174, 245
drought 180, 190, 202
dyking 29, 30, 38, 51

E

elevator companies 178, 199
equipment 8, 14, 44, 62-63, 83, 87, 88, 98, 106, 149, 150, 151, 153, 157, 162, 165, 170, 174, 175, 176, 179, 193, 200, 201, 219, 220, 230, 246, 247
experimental farms 168, 205-206, 230
export sales 141, 189, 239

F

fairs and exhibitions 44, 45, 46, 64, 87, 143-144, 169, 179, 205, 230
farming, general 242, 243
farmers' union 205
farmers' institutes 184

Index

fencing 89, 201, 206
fertilizer 116
fires 51
   forest 50, 59, 68, 78, 181
   prairie 206, 229
fishing industry 1, 2, 15
flax 1, 15, 35, 37, 58, 103, 107, 109, 128
flooding 68, 158, 165, 170, 180, 251
flour 59, 148, 214, 220, 239
food
   supply 2, 147, 177
   shortages 105
forage crops 61, 142, 204, 221, 246
frosts 12, 17, 67, 100, 117, 147, 189, 197, 201, 202, 213, 216, 235, 236
fruit
   production 59, 251
   trees 28, 37, 49, 50, 91, 99, 108, 116, 178, 226, 245, 251
fur trade 1, 4, 98, 101, 155, 165, 240

G

gardens 2, 9, 11, 25, 26, 28, 97, 98, 128, 156, 157, 158, 161, 187, 211, 212
goats 8, 9
goldrush 241
gophers 182, 203
grain storage 178
grapes 108
grass 7, 30, 242
grasshoppers 79, 91, 117, 130, 147, 162, 165, 180, 189, 190, 202, 216, 228, 230
group settlements 227
guano 89

H

hail 202
hand cultivation 36
hand tools 98, 106, 150, 157
harvesting 150
hay 9, 12, 16, 36, 159, 160, 164, 181, 212, 214, 218, 242, 244
Hebert 26
hemp 27, 35, 37, 49, 103, 109, 113, 128, 146
herbs 106, 190
Hessian fly 59, 68, 116, 117, 148
homesteads 172, 194, 196, 197
hops 251
horse ranching 225, 226

horses 8, 15, 41, 67, 78, 82, 102, 104, 107, 159, 162, 190, 248
horticultural society 49
Hudson's Bay Co. 2, 127, 156, 161, 162, 164, 169, 171, 187, 236, 240

I
incentives 34, 63, 170
Indian agriculture 4, 125, 240
Indian corn 2, 4, 9, 55, 58, 60, 99, 101, 105, 115, 126, 131, 133, 136, 161, 164, 238
insect pests 68, 117, 148, 190, 216, 229
insects 107, 147, 202
intercropping 136
interprovincial trade 177
irrigation 227, 243

K
kelp 17

L
land
  allocation 167, 172
  breaking 175, 238, 245
  claims 172
  clearing 13, 27, 29, 35, 67, 75, 84, 86, 101, 135, 136
  concessions 98
  leasing 224, 225
  ownership 171, 195, 196
  policy 12, 13, 195
  sales 241
  settlement 195
  survey 79, 171, 195, 226
legislation 22, 23, 47-49, 69-70, 92-94, 120-123, 151-152, 182-184, 206-208, 231, 251-253
livestock
  diseases 228
  general 41, 57, 59, 76-77, 82, 101, 128
  imports 18, 21, 85, 86, 89, 102, 125, 140, 144, 158, 250
  improvement 170, 224, 225, 250
  introduction 95, 102
  management 142
  rustling 229
  shows 19, 64, 65
  winter losses 250
  wintering 70
lumber industry 59, 61

Index

M

manure 17, 21, 37, 44, 108, 114, 139
market limitations 140, 141, 155
marshlands 29, 31
mice 51, 77, 78, 79, 90, 166
mildew 50, 69, 90, 148, 180
millet 60, 142
mining industry 245, 253
mission farms 126
model farms 119
mosquitoes 91

N

newspapers
  Acadian Recorder 41, 46; British Colonist 241; British Columbian 246; Calgary Tribune 230; Canadian Agriculturist 149, 152; Edmonton Bulletin 220; Free Press 177; Gazette 147; Halifax Gazette 36; Halifax Journal 62; Herald 111; Lethbridge News 225; Macleod Gazette 226, 230; Manitoba Free Press 177; Manitoban 172, 177; Norwest Farmer 203; Norwester 219; Prince Albert Times 195, 203; Quebec Gazette 111, 116; Regina Leader 194; Royal Gazette 50, 57, 61; Saskatchewan Herald 173, 201, 219; Upper Canada Gazette 145; Weekly Post 146
newspaper advertising 175
North-West Company 162, 164
North West Irrigation Act 228
North West Mounted Police 222

O

oats 2, 12, 16, 20, 55, 56, 58, 60, 79, 107, 164, 175, 244
onions 162
orchards 50, 131, 137, 138, 243
oxen 41, 56

P

pears 28, 37, 75
peas 2, 30, 37, 55, 58, 60, 79, 100, 107, 131, 132, 164, 237
pigeons 214, 215, 216
pigs 25, 26, 31, 55, 77, 82, 126, 131, 142, 158, 160, 218, 244, 249
plant diseases 181
plowing 11, 27, 157, 176, 188, 215
plowing matches 89, 143, 247
pork production 137, 146, 157
potato dry rot 69, 118

The Trail Blazers of Canadian Agriculture

potatoes 16, 17, 18, 19, 20, 36, 56, 58, 60, 72, 79, 81, 84, 104, 133, 136, 142, 161, 162, 164, 173, 188, 190, 212, 213, 215, 218, 219, 235, 236, 237, 240, 242
poultry 11, 25, 59, 83, 103, 114, 125, 213, 233, 249
predators 22, 80, 157, 160, 170, 250, 253
premiums 85
production
    costs 203, 204
    incentives 49, 113
    limitations 11
    potential 244, 245
    practices 11, 12, 17, 18, 38, 39, 42, 107, 108, 116, 137, 139, 140, 150
pumpkins 4, 237

R

railways 168, 172, 184, 191, 198, 199, 227
ranching 222, 223, 225
Red River Settlement 163, 164
roads 110
round-ups 223, 225
rust 69, 79, 90, 180
rye 2, 9, 26, 49, 53, 59, 96, 99, 100, 142

S

schools of agriculture 118
seignorial system 101, 109
settlements
    L'Anse aux Meadows 7; Battleford 193; Beaubassin 32, 54; Beauport 101; Brandon 161; Cap Rouge 95; Cap Tourmente 97; Carlton House 189; Chedabucto 30; Chignecto 31; Chilliwack 240; Chipody 56; Churchill 157; Cobequid 32; Craigflower 240; Cumberalnd House 188, 190; Cuper's Cove 8; Dunvegan 213; Fort Alexandria 238; Fort Assiniboine 218; Fort Bas de la Riviere 163; Fort Chipewayan 218; Fort Douglas 165; Fort Edmonton 212, 218; Fort Frontenac 131; Fort George 238; Fort Langley 237; Fort McLeod 212; Fort St. Charles 161; Fort St. James 235; Fort St. John 106; Fort St. Pierre 54; Fort Simpson 240; Fort Vancouver 237; Fort Vermilion 211; Fort Victoria 238; Fort Whoop-up 222; Fort William 133; Fraser Lake 234; Grand Portage 133; Halifax 35; Hudson House 187; Isle a la Crosse 191; Kinistino 193; Lac la Biche 219; Lac St. Anne Mission 219; Lake Athabasca 211; LeHave 28, 29; Lunenburg 34, 35, 36; Maugerville 57; Meliguishe 35; Minas 32; Montreal 101; New Caledonia 234; Nisqually 237; Nootka 233; Oxford House 163, 165; Port La Joye 74; Port Royal 26, 34, 35, 36; Portage la Prairie 163; Prince Albert Mission 192, 193; Quebec 95; Red River Settlement 163, 164; Sable Island 25; St. Albert 227; St. Croix Island 50; St. Peters 30; Shepody 54; Sillery 101; South Branch House 188, 193; Stuart's Lake 234, 249; Trois Rivieres 101

Index

White Earth House 213, 216; York Fort 155

Settlement leaders
  Alexander 28; Baltimore, Lord 9; Bourgeois, Jacques 54; Cartier 95; Champlain 53, 95; Cochrane, Governor 15; de la Roche 25; de Monts 25, 53, 95; Denys, Nicolas 30, 53; Frobisher 4; Guy, John 8; Henry, Alexander 162, 213; Leury, Baron de 25; La Verandrye 131, 152; Le Vallieres 54; Poutrincourt 26; Razily, Chevalier de 28; Roberval 95; Saint Pierre, Comte de 73; Selkirk, Lord 80, 166; Talon 102 Thibaudeau, Pierre 54; Wynne, Captain 9

sheep 7, 38, 41, 76, 82, 102, 107, 166

sheep ranching 225, 226

smut 90, 111, 112, 117, 149, 180, 203, 230

soil productivity 137, 138, 140

soil quality 7, 9, 29, 79, 110, 133, 163, 165, 173, 174, 190, 193, 238, 242, 254

squash 4, 125

squirrels 216, 236

stray animals 169

subsidies 34, 49

sugar beets 231

supply management 105

T

ticks 148

tillage tools 150

timber growth 134

timothy 89, 221

tobacco 4, 75, 103, 108, 146, 178

transportation 3, 141, 155, 199

turnips 16, 38, 56, 61, 75, 148, 156, 165, 187, 190, 213

U

uplands 37, 39

V

vegetable insects 216

vegetables 25, 26, 28, 37, 38, 59, 61, 97, 99, 106, 162, 163, 173, 234, 240, 242

W

weather 100, 147, 201, 250

weeds 50, 69, 92, 115, 120, 149, 181, 202, 229, 251

weight standardization 23

wheat
  general 2, 16, 17, 18, 19, 26, 28, 30, 35, 37, 39, 40, 55, 58, 59, 63, 71, 96, 100, 106, 125, 131, 135, 142, 165, 193, 197, 214, 219, 221, 237, 239

   exports 114, 177
     midge 68, 92, 117, 148
wild rice 132
winter livestock feed 138-139
winter livestock losses 228, 250
wintering conditions 70, 160, 243
wolves 22, 169